THE PLANET IN A PEBBLE

Jan Zalasiewicz is a lecturer in Geology at the University of Leicester, and was formerly with the British Geological Survey. A field geologist, palaeontologist, and stratigrapher, he teaches various aspects of geology and Earth history to undergraduate and postgraduate students, and is a researcher into fossil ecosystems and environments across over half a billion years of geological time. He has published over a hundred papers in scientific journals and is the author of *The Earth After Us: What legacy will humans leave in the rocks?* (OUP, 2008).

The Planet in a Pebble

A Journey into Earth's Deep History

JAN ZALASIEWICZ

OXFORD
UNIVERSITY PRESS

OXFORD
UNIVERSITY PRESS

Great Clarendon Street, Oxford OX2 6DP

Oxford University Press is a department of the University of Oxford.
It furthers the University's objective of excellence in research, scholarship,
and education by publishing worldwide in

Oxford New York

Auckland Cape Town Dar es Salaam Hong Kong Karachi
Kuala Lumpur Madrid Melbourne Mexico City Nairobi
New Delhi Shanghai Taipei Toronto

With offices in

Argentina Austria Brazil Chile Czech Republic France Greece
Guatemala Hungary Italy Japan Poland Portugal Singapore
South Korea Switzerland Thailand Turkey Ukraine Vietnam

Oxford is a registered trade mark of Oxford University Press
in the UK and in certain other countries

Published in the United States
by Oxford University Press Inc., New York

First published 2010
First published in paperback 2012

British Library Cataloguing in Publication Data

Data available

Library of Congress Cataloging in Publication Data

Data available

Typeset by SPI Publisher Services, Pondicherry, India
Printed in Great Britain
on acid-free paper by
Clays Ltd, Elcograf S.p.A.

ISBN 978-0-19-956970-0 (Hbk.)
ISBN 978-0-19-964569-5 (Pbk.)

10

To my colleagues who pursue the secrets of Welsh slate;
the stories in this book belong, naturally, to them.

ACKNOWLEDGEMENTS

This book began its life in a conversation with that nonpareil editor at OUP, Latha Menon, and she then delicately helped guide the resultant narrative into its current shape. Others at OUP, including Emma Marchant and Kate Farquhar-Thomson have also contributed to various stages of this book, and it has been a pleasure to work with them. Tim Colman, Jane Evans, Sarah Gabbott, Ryszard Kryza, Alex Mack, Stewart Molyneux, Melanie Leng, Derek Raine, Adrian Rushton, Andy Saunders, Sarah Sherlock, Thijs Vandenbroucke, and Dick Waters have read and considerably improved sections of this book (thanks also to Tim Colman, Sarah Gabbott, Ryszard Kryza, Rob Wilson and Thijs Vandenbroucke for supplying or helping with illustrations): to them many thanks (though errors of omission or commission in this narrative are mine alone).

Beyond that, most of this 'pebble book' forms a kind of summation—an interim summation, naturally—of work that I have been involved with, more or less tangentially, in the large part of my career devoted to untangling the intricacies of Welsh slate. In this, there is a large cast of colleagues near and far to who I owe much thanks, for educating me in this most under-appreciated type of rock (that has often had, alas, the reputation of being wet, grey, and monotonous). Much of this took place while I was working with the British Geological Survey, mapping the hills of central Wales. First of all, there were the field geologists I worked with—Dick Cave, Dick Waters, Jerry Davies, Dave Wilson, Chris Fletcher, Dave Schofield,

Tony Reedman, John Aspden, and others besides. It is hard to overstate the value of the skill and expertise that they built up as they worked year after year on long field seasons, in all weathers. The insights they developed to these difficult, perplexing rocks form, it seems to me (as someone who had a ringside seat), a true classic of British geology.

Key parts have been played too by Jane Evans and Tony Milodowski on the 'rare earth' mysteries and much else besides; Sarah Sherlock on the subtleties of rock-bound argon; Dick Merriman and Bryn Roberts on the clay minerals and micas; Keith Ball and Melanie Leng on the chemistry of these rocks; and Alex Page, deciphering both ancient life and climate from them. There is the fossil world too, that almost infinite jungle of vanished life, traversed sure-footedly by the likes of Adrian Rushton, Dennis White, Mark Williams, Barrie Rickards, David Loydell, Steve Tunnicliff, Mike Howe, Stewart Molyneaux, Phil Wilby, and Hugh Barron. Since I joined the University of Leicester, I have kept in touch with this kind of geology, mostly vicariously, as it has been pursued by such as Sarah Gabbott, Mike Branney, Mike Norry, John Hudson, Steve Temperley, Dick Aldridge, David Siveter, Andrea Snelling, Anna Chopey-Jones, Anne-Marie Fiddy, Lindsey Taylor, Bob Ganis, and Thijs Vandenbroucke. Other colleagues more widely within the Welsh orbit have included Nigel Woodcock, Denis Bates, Richard Fortey, Robin Cocks, Howard Armstrong, Derek Siveter and all those associated with those nigh-legendary (if tiny) institutions, the LRG (the Ludlow Research Group) and BIG G (the British and Irish Graptolite Group), and more lately the Welsh Basin Group. And, going back to my own ancient history, John Norton and then Harry Whittington played key roles in starting me on this path.

To these and to yet others I am truly in debt: most of the stories in the following pages belong to—and hence the book is dedicated to—these people.

Pebbles and rocks, though, go back a long way in my life. My parents and sister patiently tolerated—indeed supported and encouraged—my early excavation of such things, despite the large volume of rock that I insisted on carrying into a small house. More recently, my wife Kasia and son Mateusz have borne the brunt of the time stolen to fashion and refashion the words in these pages (not to mention the long weeks and months when I have been in the field amid the Welsh hills). To these also I am eternally grateful.

CONTENTS

LIST OF PLATES

Plate 1: **a** Welsh slate with more resistant sandstone strata 'ribs', surrounded by pebbles formed from its destruction by the sea. Clarach Bay, Wales.

b The underside of a sandstone bed showing flute casts—the sediment-infilled erosional scours formed by vortices within a turbidity current.

Plate 2: **a&b** The two main types of Silurian sea floor. On the left (**2a**), sandwiched between homogeneous grey, rapidly deposited turbidite muds is a unit of dark, organic-rich finely laminated mudstones laid down on an anoxic sea floor. On the right (**2b**)—its *alter ego*: a pale mudstone layer with conspicuous dark burrows, representing an oxygenated sea floor, colonized by worms and other multicellular creatures.

c-f A variety of fossilized graptolites, preserved in various combinations of shiny black carbon, pale golden pyrite and orange to brown iron oxides. The conspicuous pale patch surrounding the graptolite in 2f is from chemical alteration of the mudrock around the fossil.

Plate 3: **a-c** Monazite nodules. Optical microscope view (**3a**) of a thin section of Welsh slate, showing the monazites as three black oval patches in the dark laminated layer. More detailed views (**b-c**) taken using a scanning electron microscope; the monazite shows up so brightly because its dense atomic structure reflects more electrons than does the surrounding rock.

3d A fold in Welsh mudrock and (**3e**) the kind of tectonic (slaty) cleavage produced in the mudrock as a result of the immense pressures exerted on the rock—the cleavage is the near-vertical structure, cutting across the fainter sedimentary stratification that is near-horizontal.

Plate 4: **a** A tectonically thickened barrel-like mica (less than a millimetre across), as seen by scanning electron microscope.

4b A matchstick-sized graptolite surrounded by a 'halo' of fibrous mica (stained orange by iron oxides), formed around these fossils during mountain building.

4c-e Polished surfaces of metal ores (pyrite, chalcopyrite, galena) as seen using a reflected-light microscope (pictures taken by Tim Colman).

PROLOGUE

It is just an ordinary pebble. One of millions that wash backwards and forwards on the world's shorelines, or pile up on riverbanks or perhaps line your garden path. Yet that pebble, like its myriad kin, is a capsule of stories. There are countless stories packed tightly within that pebble, more tightly than sardines in the most ergonomic of tins.

The size of this story-capsule is deceptive. These stories are gigantic, and reach realms well beyond human experience, even beyond human imagination. They extend back to the Earth's formation—and then yet farther back, to the births and deaths of ancient stars. Something of the Earth's future, too, may be glimpsed beneath its smooth contours. Battle, murder, and sudden death are there, and ages of serenity too, and molecular sleights of hand that would make a magician gasp; there are extremes of cold in those stories, and also temperatures that far surpass those at the heart of our Sun.

The human mind loves to make up stories. It can take any everyday object and weave around it tales of elves and princesses, of witches, goblins and vanished empires. We are born storytellers, and were likely so even in the mists of human prehistory.

Today, though, stories can be discovered that are no less evocative, but they catch glimpses of the reality of this Earth and the cosmos that surrounds it. They concern not what might be simply imagined of the Earth's past, but what can be *deduced* from the evidence of what can be seen, measured, detected, analysed, compared.

The stories woven thus are as beautiful, and surprising, as the ones plucked from imagination alone. But they are held, constrained by that evidence, evidence that is often won through inventiveness, patience and sheer persistence—and sometimes cunning, and luck, too. These stories cannot contain anything that can be shown to be untrue, at the time of telling. This is not to say that the pictures we create of the world's past are true, necessarily. But, they are the most reasonable interpretations that we can manage, pictures that are more or less in focus, better or worse approximations, depending on the nature of that evidence.

We are not just born storytellers. We are technically ingenious, and have devised means to detect, to analyse, to measure that seem, still, to me, to verge on the miraculous. And from a combination of narrative and technical wizardry, the distant past can be recreated, brought to life. One can take a single pebble, and enlarge it to the size of a mountain (or greater) by means of our amazing machines, and pursue an almost infinite number of paths within it, each reaching into a different corner of some vanished earthly—or unearthly—environment.

Most of these stories lie buried in the forests of printed paper in our libraries—or now increasingly form part of the electronic web of information that surrounds us all.

Here are just 13 of the paths that can be taken through that single pebble. Why 13? Well, it seems about the right number to connect the stories into something that is, if not a cycle, at least a chain that goes from distant beginnings into a far future. And I'm a British geologist, and the beginnings of organized geology in Britain (and, one might say, in the world) took place on Friday 13th November 1807 in the Freemasons' Tavern in Great Queen Street, Lincoln's Inn Fields, as the Geological Society of London was born. There were 13 original members of that Society. It's a kind of reverse superstition, if you like. Thirteen was a lucky number in the growth of the science.

And the pebble—which one—and from where? What pebble would I take with me on a desert island, to be a decoration and keepsake among the ornate shells and carved wood and coconut shells? It would be a pebble of grey slate from a Welsh beach—perhaps from somewhere like Aberystwyth, or Clarach, or Borth on the west Wales coast—or it might be from inland, from the gravel-lined banks of rivers like the Ystwyth or the Rheidol or the Claerwen. I've spent more than half my lifetime amid the rocks of the Welsh crags and cliffs and hillsides, trying to interpret some of the distant histories that they contain. These beach and river gravels are mostly made up of the rocks eroded from those cliffs and crags. They are often disc-shaped, and will fit nicely into the palm of your hand—or, thrown flat across the waves, will bounce and skim over the water's surface.

On a wet day, or dipped into the water at the sea's edge, the colours show up; shades of grey and blue-grey, often in delicate stripes. Some are cross-cut by thin crazed white lines, like small frozen lightening tracks, or here and there have flecks of gold or red. The shapes and textures and colours all have their own stories.

Time to select a nice one—maybe just that one ... there!—and start. From the beginning.

Stardust

What is a pebble? It is a wave-smoothed piece of rock, and a complex mineral framework, and a tiny part of a beach, and a capsule of history too. All these guises have their own stories, and these we shall come to. But from yet another viewpoint the pebble is a collection of atoms of different kinds—of many, *many* atoms—and that might be the best way to start. Considering it at this level, it is a little like taking the equivalent of a large sack of mixed sweets and separating them out into their different types. How big a sack, though? Or, to put it another way, how many atoms in our pebble?

There is a simple formula for estimating the number of atoms in a piece of anything. The basic idea was first glimpsed by Amadeo Avogadro, Count of Quereta and Cerreto in Piedmont, now Italy: scholar, savant and teacher (though his teaching was briefly interrupted because of his revolutionary and republican leanings—a little impolitic when the king lives nearby). Avogadro was interested in how the particles (atoms, molecules) in matter are related to the volume and mass of that matter. Years later, his early studies were refined by other scientists and the upshot, a century or so later, came to be called Avogadro's constant. Thus, in what is called the mole of any element there are a little over 600,000 million million million—or, to put it more briefly, 6.022×10^{23}—atoms. A mole here is not a small furry burrowing quadruped, or a minor skin blemish, but the atomic weight of any element expressed in grams. For oxygen a mole would therefore be 16 grams, as 16 is its atomic weight, an oxygen atom having a total of 16 protons and neutrons in its nucleus.

The kitchen scales tell us that our pebble weighs some 50 grams. About half of it is made up of oxygen, and much of the rest is silicon (atomic weight 28) and aluminium (atomic weight 27) with a scattering of other elements, most somewhat heavier. A judiciously averaged atomic weight might therefore reasonably be something like 25. Our 50-gram pebble therefore contains of the order of one million million million million atoms (or thereabouts). If our atoms were sweets—those nice wrapped ones, where every centre is different—then the sack we would need to fit them in would be about the size of the Moon.[1] This is a measure of the enormity of the submicroscopic world that everywhere surrounds us. William Blake famously saw the

[1] One can work it out—with the sweets, that is. The Moon's volume is about 20 thousand million cubic kilometres (as worked out by some patient astronomer), each cubic kilometre containing a million million litres—and a litre is the capacity of our kitchen measuring jar, which could fit 50 sweets inside it. Those numbers, multiplied together, come to about one million million million million, or close to the same number as the atoms in our pebble.

world in a grain of sand—his intuition was out by only an order of magnitude or so.

We therefore have a superabundance of atoms. But atoms of what kind? Let us take an inventory. There are machines today that count atoms. The very expensive, de luxe models can do this atom by atom, almost. But even a standard laboratory atom-counter—more exactly an X-ray fluorescence spectrometer—will give us a reasonable first estimate. It does not count atoms individually, but it measures the proportions of different types of them, that is, of the different elements. To use it in practice, one would first have to sacrifice the pebble. That would never do, so early in our story. So we will sacrifice one of its neighbours instead. The results will serve us just as well. For purposes such as this, slate (at least from the same cliff-stratum, on the same beach) is slate is slate.

The sacrificial pebble is crushed, pressed into a pellet, or melted into a rock-glass bead. Then, a beam of high-energy X-rays is fired at the sample, and these X-rays knock some electrons out of their orbits. Other electrons, falling in to take their place, give off radiation (photons) as they fall, photons whose pattern of energy is characteristic of, and can thus 'fingerprint', each element. These photons, and their energy levels, are detected and measured, very precisely, and so the proportions of elements in the sample can be measured, to within a couple of per cent or so.

We will then, like atomic accountants, have one measure of the pebble, in approximate percentages. On the balance sheet, oxygen appears as the most common type of atom, by far. At about half of the pebble by mass, it makes up a greater proportion of this piece of rock than it does of the atmosphere. A surprise, perhaps? Well, the pebble cannot be used as an oxygen bottle. More's the pity, for if one could do this, then astronauts could breathe easy on the Moon and on Mars. The oxygen is locked away, bound up tightly with silicon and aluminium to form the mineral framework of the rock, as silicates of various kinds. We will say a lot more about *those* anon. But for now suffice to say that in that mineral skeleton there is also room for the other main components, weighing in at a few per cent each: iron and magnesium, potassium and sodium, and titanium too. Then, at a bit less than one per cent, there is calcium and manganese and phosphorus. Here also one might include that proportion of the rock that the analyst calls 'loss on ignition', which is what is driven off when the rock powder is baked in an oven; it is mainly water, the rock's main store of hydrogen (bound up—inevitably—with oxygen), together with a soupçon of carbon.

Now we are into the rarities, those elements measured in parts per million, or simply 'ppm'. Rare maybe, but diverse certainly; there is a whole array of them. Some are familiar, weighing in at a few hundreds to a few tens of ppm—vanadium, chromium, copper, zinc, lead, barium; others are a little less well known—rubidium, strontium, yttrium, cerium, lanthanum, niobium. One or two are a little sinister—arsenic, for instance. Some of the atoms here are radioactive. These are the ones where the proportion of neutrons and protons in the nucleus is not quite right, where there is an inner instability, a nuclear tension. This will cause that atom eventually (after microseconds, or billions of years, depending on the type of atom) to snap apart, to cleave into smaller atoms, and also into the shrapnel (high-energy electrons, or gamma rays, or bound pairs of neutrons and protons) that we fear as radiation. There is for instance uranium, which comes in no less than sixteen different types, or *isotopes* (all of them radioactive), each with a different number of neutrons to accompany the constant 92 protons. There is also thorium in this category, and some of the isotopes of potassium and samarium and rubidium.

There are rarer elements still in the pebble, present in parts per billion. To detect these we need even more sensitive equipment—mass spectrometers, in which an accelerated plasma of pebble material is nudged by magnets so that streams of charged atoms, or ions, are carefully sifted by weight as they are sent flying into carefully placed detectors. Here are the elements of which poets and pirates speak—gold and silver and platinum. Vanishingly rare—and yet there will be many, many millions of each of those types of atom in the pebble. In an almost infinite jungle, even the rarest orchid is abundant. At this level, rather than saying which elements are present in our pebble, one might try to say which are *not*.

It is hard to find one such. Iridium is often cited as the rarest of the elements in the Earth's crust, with an average abundance of one part in a billion. Well, with our million billion billion atoms, that means, astonishingly, that there must be a million billion atoms of iridium in the pebble, each atom being almost—but not quite—infinitely diluted among the vast abundances of its more common neighbours.

Iridium at least is stable. There are those elements that, like uranium, are not. Of these, among the most fugitive and fleeting is the highly unstable promethium, a by-product of the decay of uranium. It has a half-life of a little less than seventeen years. And yet, with uranium present at a few parts per million, there will be thousands of prometheum atoms winking in and out, invisibly and undetectably, within the enormous atomic vault of the pebble.

And then, there is the shadowland beyond uranium in the periodic table, the ghostly world of those elements too heavy to be stable, like uranium, but with half-lives far shorter. Most we know as artificial: plutonium and americium and einsteinium and mendelevium—and now copernicium, if the name is ratified (that is, if it is not found to mean something very rude in another language). With some of these, like mendelevium, we may have true rarity—perhaps the lack of even a single atom within the boundless spaces of the pebble. But others of these types of superheavy atoms, like plutonium, have now been produced by *us* in nuclear reactors in considerable quantity—by the kilogram, in fact. Released into the realm of wind and waves and running water, there is now more than likely a sprinkling of such elements adhering to the surface of our pebble, and to that of every other pebble on the beach (and all pebbles on all beaches on Earth today).

Our pebble is truly a microcosm of the Universe. And those atoms of the pebble, the hundred-odd varieties of them, have a place and a history within the Universe. One needs, though, to separate place from history. For the pebble, while nicely representative of the Earth's upper crust, is—along with the rest of the Earth and the inner planets—cosmically unrepresentative. The Universe is mostly hydrogen and helium[2] (though that is changing, very slowly). The pebble—and the Earth it rests on—is oxygen and silicon and iron and suchlike heavy elements; its hydrogen content is mostly within water and its helium

[2] At least as far as conventional matter is concerned. The Universe, in bulk, behaves as though most of it is made of some mysterious other things: one that we call 'dark energy', that is pushing it apart on one scale, and something else that is called 'dark matter' that is pulling parts of it together on another. The pebble at this level is insignificant, as are (almost) all the rest of the familiar atoms in the universe.

content is negligible. So this cosmic otherness of the pebble needs explaining. And before even that, there is the question of the ultimate origin of the atoms within our fragment of slate. For their birth has been extraordinary, and their journey has been unimaginably long, and they have passed through realms that no spaceship will ever penetrate.

FIRST JOURNEY

Where might this first journey through our pebble of slate begin? It must begin at the only beginning that we are aware of, when all the stuff of the pebble was created—as was the stuff of *us*, who are now considering that pebble, and of the Earth that we live on, and of everything that we see as we gaze into the night sky above us.

A singular moment, this, in our history, an extraordinary instant in which something was apparently created from nothing, and that 'something' then spread outwards from its tiny origin to become all of the Universe. It has been christened the Big Bang, and it happened about 13.7 billion years ago.

The pebble, in this respect, is as deep a mystery as is everything else in the Universe. How *did* the matter of that pebble, and of the Welsh hills it was torn from, and of the world it sits atop—and of the Solar System, and of the Milky Way, and of countless galaxies near and far—manage to unpack itself from a point: a 'singularity', as many think, of no size at all? And how could it unpack itself, or rather inflate, as the process is more generally termed, at such scarcely believable speed? It went—it is thought—from a standing start at microscopic size to bigger than a galaxy in less than a second, seemingly far outpacing light itself.

The atoms of our pebble, now so poised and tranquil in their mineral frameworks, are representatives of the few survivors of a battle of subatomic particles, quarks and leptons and their kin, that took place soon after that singular event. It was a battle of great intensity, carried out at temperatures—of many trillions of trillions of degrees—that were never repeated. Most of these particles did not survive, as particle and antiparticle annihilated each other, releasing energy to further fuel the cosmic fire.

Did regions of antimatter escape, to ultimately form anti-pebbles on the anti-shore of an anti-planet in some far-off anti-galaxy? Such ideas inhabited the science fiction of my youth, and astronomers have searched the night sky for the telltale flashes of light that would be emitted as antimatter galaxies encountered galaxies made of normal matter, with opposing particles transformed into pure energy as they met. Alas, no such flashes have been found. Either matter won out convincingly, in those furious first few minutes of the Universe, or else those anti-pebbles lie on shores so distant from our own, as to be out of sight of even the most powerful telescopes.

In whichever version, the stuff of our pebble atoms began to take on part of their more familiar aspect less than a second or so into the universe. Protons and neutrons appeared within a universe that, albeit still cooling and expanding furiously, remained *in toto* a densely packed fusion furnace in which the nuclei of future atoms could emerge, at temperatures falling fast towards a billion degrees. A proton, by itself, is a simple hydrogen nucleus, and that was—and remains today—the most common building brick in the visible Universe. Have a neutron smash into a proton and, combined, these two become a deuterium nucleus; hammer in another proton (and optionally one more neutron) and it becomes a helium nucleus. A good deal of helium originated this way—perhaps making up a quarter of the matter then formed. In a tiny, tiny fraction of a per cent of these events a further proton was successfully added, and so there appeared a minute sprinkling of the nuclei of some lithium-to-be.

Three minutes had passed. And that, for a long time, was it—at least as regards the pebble matter. Our ingredients are now a plasma of nuclei and electrons, still speeding outwards, in a universe that is opaque as pea soup, as photons cannon and rebound continually amid the mass of milling particles.

A quarter of a million years passed. And then, there was light. The Universe became transparent. Electrons became captured to form neutral atoms, and photons, that were finally able to travel through the thinning particles, lit up the Universe. Our pebble ingredients were now bathed in the universal glow given off by a Universe that was now at a mere 3000 degrees Celsius. Astronomers can still detect that light as a faint afterglow, the cosmic microwave background, visible in every direction, everywhere they look in the sky, though this radiation has, over the intervening 13.7 billion years, stretched into microwaves and cooled to three degrees above absolute zero.

It was a large step towards a familiar normality. The temperature of the particles, still speeding outwards, had dropped to the level at which electrons could be captured by those collision-generated nuclei, and settle into the orbits that they would continue to occupy, virtually eternally (though of course being prone to migrate promiscuously from atom to atom in that phenomenon that we know as electricity). Atoms had appeared for the first time: of hydrogen, of helium, and of that tiny sprinkling of lithium.

And so were born a few of the atoms of our pebble. The hydrogen there, bound into the water molecules present in the slate-matter, thus has the most ancient pedigree of all. It does not, of course, show signs of its immense antiquity. The proton at its heart is as good as new, the sole electron as tireless in its orbit as when it first dropped into place.

Most of our other pebble atoms had to wait much longer for their birth. The raw materials were those outrushing clouds of hydrogen and helium filling the early expanding Universe. A further alchemy was needed to build these larger atoms. But the furnaces in which such construction could take place could not be built then, not quite yet.

For they were rushing outwards, these atoms, within a gathering darkness, a darkness that was growing as the temperatures continued to drop, and as the glow from that first light outburst faded. Now came a time of gas clouds, and the development of universal *unevenness*. Because the further development of our story—and indeed the eventual emergence of the chroniclers of that story—hinged upon imperfections of the Big Bang.

Had that initial explosion been perfectly regular, spreading its products perfectly evenly into the expanding Universe, then the entire history of the Universe would have been that of an ever-thinner and ever-cooler dispersal of lone atoms, becoming ever more solitary as they sped outwards. But that explosion showed fluctuations, irregularities, which can be seen even now as regional variations in the strength of that primordial, almost-cooled universal afterglow that pervades outer space.

Where the gas clouds were thicker, gravity (another of the inventions of the Big Bang, along with the rest of the physical laws and forces) came into play. The hydrogen and helium atoms, and those few lithium atom pioneers, slowly began to be drawn together in that long Dark Age of the early Universe.

In the more massive clouds, gas was falling into the ever-denser cores of massive balls of gas. Unable to go further, the gas decelerated. As it was compressed, its energy of motion turned to heat. When, in one of those balls of gas, the heat thus generated first built up to 100 million degrees or so, nuclear reactions began, and a pinpoint of new light began to shine in the universe. The first star-furnace was switched on, and the first star was born.

It was a necessary prerequisite to creating the bulk of our pebble stuff, but that creation would have to wait a little longer. For, the core of a just-lit star is still an insufficient nursery to grow atoms larger than helium, and 100 million degrees Celsius is still, by far, too cool a forge for our needs. At that temperature, one speeding hydrogen nucleus has sufficient energy to overcome the repulsion that normally keeps naked protons apart, and it can collide with another nucleus. Once they are pressed close enough to each other, the strong nuclear force can lock the two nuclei together. Four protons then combine in a series of reactions, creating a helium nucleus in which a tiny part of the mass of the whole has been lost. This lost mass has been converted to energy, and into quite enormous amounts of energy, as Einstein showed in his discovery of the world's most famous equation: $E = mc^2$. This simply states that the energy generated is equal to the lost mass multiplied by the speed of light, and then multiplied *again* by the speed of light. The speed of light multiplied by itself is 9×10^{16} (metres per second)^2. This figure is enormous. It is a measure of the fusion power that drives the star.

But, to assemble more protons and neutrons together into bigger atom-units takes more intense conditions still. Those of a small star like our own will not do. The Sun is, on the galactic

11

scale, an ordinary slow burner, as are most stars. It has steadily burnt for a little over 4.5 billion years, slowly converting its hydrogen fuel to helium. It has incubated life on at least one of its planets, and warmed it just sufficiently to allow it to flourish, and it will burn like this for another 5 billion years or so.

It takes a big star, ten times the size of our Sun or more, to act as forge for the kind of elements that can make a rocky planet—and a pebble. In those giant stars the furnaces are bigger, and burn hotter and faster. These giant stars use up their fuel voraciously, and can race through even their capacious supplies of hydrogen in a mere few million years. When the nuclear fires begin to dim, the interior of the star, no longer able to resist the crushing gravitational forces, starts to collapse. But the collapse, in turn, generates heat, from the friction of the tightly packed infalling atoms. Heat is simply a measure of the speed of moving atoms and, at some 200 million degrees, helium atoms are moving fast enough for their nuclei to fuse, thus reigniting the nuclear furnace. Carbon, then, is born, and life—as we know it, at least—in the Universe has been made possible.

But in this fast-evolving star, helium, too, is quickly exhausted, and the next phase of collapse and reignition starts. Carbon itself ignites at 800 million degrees to give birth to yet more elements: oxygen, neon, sodium, magnesium. And so it goes on, as neon, and then oxygen, and then silicon successively provide fuel for the furnace—and feedstock for new elements—at temperatures that are now reaching 3000 million degrees at the core of the star.

There is a limit to this stellar alchemy. Hammering together nuclei and building bigger atoms releases energy until the atoms are as big as those of iron (that have a total of 56 protons and neutrons in the nucleus). Atoms that are bigger than those of iron cannot be easily made by this process, for these larger-scale fusions *absorb* energy, quenching the nuclear fires, and so the furnaces die. Then, the stage is set for the final flourish, an awesome denouement that provides the ultimate atomic factory, where the larger atoms of our pebble are created: the copper and zinc, the arsenic and lead and lanthanum, and the gold and silver and platinum too. The star thus dies— gloriously, and terribly, and productively.

There takes place a spectacle such as the one that burst some 7300 years ago, out of a far corner of the Milky Way galaxy. Then, 6300 years later, the leading edge of the outburst of light and energy reached this planet. For three weeks, it became, bar the Sun, the brightest object in the sky, night and day. It was seen— and recorded—by curious and awestruck watchers of the sky in Arabia, China, Japan, in native North American homelands too, maybe, and perhaps even in Irish monasteries (as an event transmuted, later, into a story of the Antichrist). It then faded. Two years later, it was lost from view even in the night sky. But train a telescope on the region today, and the star-wreckage of the mighty Crab Nebula appears. It is the remains of a supernova.

Such a star-death is exceptionally energetic, even by the standards of this violent Universe of ours. Were our Sun to become one (and it can't, thank goodness, being too puny), then the sky would fill with the light of a sun that had suddenly become 10 billion times brighter. The flood of energy, briefly equal to that shining from every other star in the galaxy, would vaporize any Earth-bound observer even before the event could register in eye and brain. Then, it would take only a few days to sand-blast away the Earth itself. But amidst such a spectacle of power, we are concerned with one thing: it is the only anvil fit to forge the final atoms of our pebble.

The supernova starts with the final collapse of a giant star that has used up all its fuel. As the star implodes, its interior briefly recreates the moments after the Big Bang. A blizzard of neutrons is released that, hammering into the tightly packed atomic nuclei, lodge within them to drive up the size of those atoms. In that short-lived maelstrom at the dying star's heart, where notions of temperature and pressure are now beyond any human comprehension, all the remaining elements are created, up to lead, uranium and beyond. Not all of these new atoms are built to last, though. Jerry-built as they are by that chaotic hurricane of subatomic particles, they may be constructed with too many, or too few neutrons relative to protons. These unstable atoms may fall apart within a fraction of a second, or days, or years, or aeons, after they were formed, their degree of stability depending on the exact manner of their (mis) construction.

Their sojourn within the star is brief. The catastrophic implosion rebounds, and most of the star is turned inside out in the ensuing explosion of the supernova. The elements that have been created are flung with the stellar debris into outer space, and they begin journeys across immense stretches of interstellar space. They condense into the first minerals, becoming stardust: the building blocks of new stars and planets, including our own, and the raw ingredients of us—and of the pebble.

We humans can see stardust today, from afar. Our galaxy contains clouds of dust that, where they swirl thickly enough—like the much-photographed Horsehead Nebula—can block out starlight. It envelops young stars that are close enough to us to examine individually, and here the light shining through the dust can be analysed by telescope and spectroscope. The dust, thus interrogated, includes familiar minerals: silicates of iron and magnesium such as olivine and pyroxene, as sand-sized particles. There is carbon, too, as tiny specks and as longer hairs called 'graphite whiskers' that give their own telltale spectral signal to the astronomers, and also as microscopic diamonds. Humans have caught some of these interstellar particles: the spaceship of the *Stardust* mission used a special gel-coated collector to catch a few specks out of a stream of dust particles thought to originate from beyond our Solar System. And tiny specks of dust from ancient, distant star systems have, by meticulous chemical analysis, been found within meteorites. These grains—oxides and silicates—are telltale aliens, as the proportions of different types of individual atoms (isotopes) are wildly different from anything native to this Solar System.

From how many supernovae have the atoms in our pebble been derived? And how far did they travel across the vastness of interstellar space, before they arrived in the cloud of dust and gas that was to become our Solar System? Many of the pebble atoms probably passed through previous solar systems, were engulfed in growing stars, and then were expelled once more (together with other more newly minted atoms) in the death throes of those stars.

We can get a sense of this history, through the telescopes (and particularly the remarkable Hubble telescope) with which astronomers peer farther and farther into space and simultaneously deeper and deeper into time. They can now capture light that has been travelling for over 12 billion years: light that records the birth (it is thought) of some of the first stars, less than a billion years after the Big Bang.

It was a time of giants, for these early stars were huge. They needed to be, for the primordial hydrogen and helium gas was hotter, and so at higher pressures than today. Only the largest clouds possessed the gravitational force to overcome this thermally generated pressure, and so only giant stars—a few hundred times the size of our Sun—could form. These lived fast, and died young, and their deaths drove the cosmic factory of the chemical elements. The light of the most distant quasar known illuminates dust containing carbon, oxygen, iron, already forming clouds less than a billion years after the Big Bang.

The paths of our pebble atoms were therefore long and mysterious, and in detail probably quite unknowable—with one exception. Stellar violence immediately preceded—and perhaps acted as necessary midwife to—our Solar System. Somewhere close to the cloud of gas and dust that was to form our own distinct corner of the galaxy, a supernova exploded, seeding this cloud with new matter. This matter included very short-lived and highly radioactive elements, such as an isotope of aluminium with one less neutron than usual. These unstable elements decayed away in a brief few million years after their creation. They have left a telltale imprint on the remains of the earliest mineral matter—the meteorites—in our Solar System, and the debris left after this early decay will be within the pebble too.

The nearby supernova that erupted and seeded the proto-Solar System cloud with these highly radioactive elements, might have played a greater role still. The shock waves from this supernova perhaps provided the 'push' to trigger the final gathering and collapse of the cloud of dust and gas that, under the gravitational attraction of its own mass, became the Solar System. Thus, out of cosmic violence might have been born our planetary home, and the Sun that warms it.

This collapsing cloud of gas and dust evolved into the central ball in which the thermonuclear fires were to be ignited—the Sun—and the surrounding disc of gas and dust that was to be the nursery of the future material, including the pebble stuff. This disc produced the known planets, and the many asteroids and comets that extend past the outermost known planet, and on, on, into the vast icy distances of the Oort Cloud, where invisible, dark representatives of our Solar System extend up to, perhaps, a light-year from the Sun, a quarter of the distance to Proxima Centauri, our nearest star-neighbour.

In the birth of this new star system, separation took place of the elements, and the segregation of those cosmic rarities—of silicon and aluminium, iron, magnesium and oxygen (that make up but a one-thousandth part of the stuff of the Universe), from the universally commonplace hydrogen and helium. The dust particles containing the pebble atoms now whirled around the Sun, in that zone where the rocky planets were to emerge.

This was a zone of energy and violence, and of mystery too. The dust particles turned into a rain of molten rock as they began their transformation into a planet. Break open a stony meteorite today and you may see the evidence, even with a simple magnifying glass. These meteorites are fragments of the original rubble that did not get swept up into the construction of planets—or perhaps started the construction process before being shattered into fragments once more. The meteorite, as you observe it under the lens, is made of thousands of tiny spheres stuck together, a little like one might imagine fossil caviar to be. These are chondrules, and they are tiny frozen melt rock-droplets that were somehow flash-heated so strongly—up to 1500 degrees Celsius—that not only was water evaporated from them, but so also were elements such as potassium and magnesium and iron. The chondrules formed dense clouds of incandescent magma-rain, hundreds to thousands of kilometres across.

What heated them so strongly? It was not the Sun's heat, for this barely ignited star was still too faint and too distant. Nor lightning bolts arcing through the dust-clouds—the melting was too thorough and widespread. Heating by the short-lived highly radioactive elements produced by the 'trigger' supernova may have played a part, but by itself would not have been enough. So what was the main heat source? It may, some think, have been mighty shock waves racing across the entire solar nebula, driven by the gravitational tug-of-war between the Sun itself and the mass of material in the debris ring around it. If earthly astronauts ever travel to experience the birth of a distant star system, they had better beware: these are perilous places.

Perilous, but fertile too. The turbulent and fiery chondrule-clouds may have been dense enough to possess their own gravitational fields, allowing them to collapse and aggregate into asteroids and yet larger planetesimals, several tens of kilometres across—the seeds of future planets. Occupying the same orbital plane around the Sun, these collided and shattered each other into smithereens, or cannibalized one another, and grew.

Planet construction is a hasty business. It would have been a brief sojourn for our pebble atoms within this energetic, whirling disc of dust, rock fragments and melt droplets. It took only a few million years of collisions within this maelstrom for most of our future pebble atoms to be safely gathered into one of the large new planets—the third (or perhaps, then, the fourth) out from the Sun.

A little over 4.5 billion years ago, most (but, significantly, not all) of our pebble constituents had drawn so close together, from their far-flung origins, as to be spread somewhere within a ball of rock that is some ten thousand kilometres in diameter (and thus a few per cent smaller than the Earth as it is now). And not just anywhere within it. They (particularly the atoms of iron and nickel that are among the ingredients of the future pebble) survived another separation process. They escaped being gathered as metallic melt droplets that, by gravity, trickled deeper and deeper, thousands of kilometres downwards, to form the Earth's metallic, nickel–iron core. This process has been called the 'iron catastrophe', the heat released creating a magma ocean above. There would have been no living casualties, though, for this was still a lifeless Earth.

The nickel and iron atoms (and some sulphur too) in our pebble might be termed inanimate survivors of this great separation—as would those rare atoms of iridium and gold, for most of those were dragged down into the core too, way out of reach of any miner's pick.

Within that great outer shell of the young Earth, where will the atoms of our future pebble have been? Probably still widely scattered, dispersed among countless mineral grains, somewhere within that nearly 3000 km-thick mixture of rock and magma that we call the mantle. They would have been dispersed among currents of mantle rock that had begun to circulate the four quarters of the Earth, shallow and deep. They were closer together than at any time before, yet still impossibly dilute within the mass of mantle material.

There were some pebble atoms still beyond our planet, too. It is now somewhere between 5 and 20 million years after the Earth has accreted from the coalesced space-rubble and planetesimals. The pebble ingredients are, very suddenly, very violently, about to be stirred and reshaped by an event which will provide for them a fundamental rearrangement, just as this event will set the course for the Earth's future evolution. In terms of violence, and energy, this pales into insignificance beside, say, those stellar explosions that forged the bulk of the pebble atoms—not to mention the unimaginable conditions of the Big Bang.

Nevertheless, this event would take centre stage in any science-fiction blockbuster film. The last main cohort of our pebble atoms is approaching the Earth, at thousands of kilo-metres an hour. These come from another planet, a planet that is doomed, for it is on a collision course with Earth. Theia is coming.

From the depths
of the Earth

The planet Theia had, like the Earth, formed early, from the mass of dust and rock-melt droplets of the accretionary disc. Theia is calculated to have been about the size of Mars, yet it was to have nothing like that planet's longevity. Its orbit was close enough to that of Earth for collision to be certain, sooner or later.

The two planets came together at something of the order of 40,000 kilometres per hour. Theia lost its separate identity over a few tumultuous minutes, and the Earth was smashed, like a grapefruit hit hard with a hammer. In that conflagration the material of the two planets, having instantly converted into boiling magma and vapour, simply merged. Theia's core sank to join the Earth's. Some of the outer layer of both planets splashed out into a cloud of plasma that encircled the suddenly re-formed Earth, and that cloud condensed to form a new companion to our planet—the Moon.

It is a fine story, this, of the Moon's creation through a spectacular planetary collision. It is likely true, too: though it is not certainly so, simply being the hypothesis that now best explains the character of our Earth and its satellite. Like many such stories in science, it is currently the one that best fits the evidence. It has been calculated, on the basis of these two bodies' mass, momentum and orbit, that it would have been extraordinarily difficult for the Earth to have captured intact a stray planetary body of this size. However, holding onto a mass of ejecta flung out by impact, kept in balance by the twin, opposing forces of gravity and centrifugal force, is a more plausible means of having formed the Moon.

More, this concept explains the remarkable isotopic similarity of these two bodies, generated by the intense mixing associated with impact. Mars, by contrast, has very different proportions of, say, the different isotopes of oxygen, because it formed in a different part of the Solar System, where the atoms of the original accretionary disk had been shuffled into different combinations. Thus, using a mass spectrometer to measure the proportions of isotopes, one can easily distinguish the few Mars-derived meteorites that have been found from all the others, just as one can pick out an orange from a bag of apples. The impact hypothesis can also explain the remarkable dryness of the Moon: its volatiles (such as water) were lost from the superheated impact plasma before it coalesced into its new shape.

A few years after this hypothesized impact, both Earth and its new satellite, the Moon, would have been molten and glowing. The Moon was closer to Earth than today; to any (entirely imaginary) Earth-placed observer it would have seemed to be twice or more the size that we see today. The Earth, then, restarted its history.

There was no question of anything so obvious as a crater surviving on our planet, as a physical record of this event. Following the impact, the Earth's outer layers would have become a magma ocean that was thousands of kilometres deep. As this slowly solidified, the planet acquired a brand new surface.[3]

The material of the Moon had now been forever torn away from that of the Earth. Among the atoms of the latter, one tiny subpopulation was to gather, some 3 billion years later, into our chosen pebble. Wherever these had been prior to the impact, their distribution will have been profoundly altered by the impact-related mixing, and by the great number of Theia-atoms—some undoubtedly present in our pebble—that are now inextricably mixed in with those of Earth.

IN THE DEPTHS

The Earth's magma ocean took many millions of years to cool. Eventually, the surface began to solidify sufficiently to form patches of a solid outer crust. These—the earliest forerunners of the continents—were slowly dragged across the Earth's surface by convection currents in the molten rock beneath. At depth, the magma ocean also began to cool and to crystallize,

[3] Mars, for comparison, may bear the traces of an impact almost as great, for its division into a rocky 'highland' southern hemisphere and its smooth flat 'lowland' northern hemisphere has been explained—controversially, still—in terms of a similar, gigantic impact in the early Solar System. This supposed impact was not quite great enough to melt and refashion the planet, but it left—according to some—an impact crater that, effectively covering half the planet, still represents its dominant geographical pattern today.

to form the (mostly) solid rock of the mantle, currents of which nevertheless continued to circulate.

These Earth-currents began to settle into the pattern that is still active today, driving the motion of continents via the mechanism of plate tectonics—a uniquely terrestrial pattern within the Solar System. Upwelling mantle currents diverged, to produce spreading areas of ocean crust at the surface, that then eventually sank downwards back into Earth's deep interior, the sinking plates in turn helping to drive the mantle currents. Up above, as the surface cooled, the rains started. Water—partly derived from volcanic degassing of the Earth's interior and partly from the arrival of many ice-rich comets— condensed out to form the first of the Earth's oceans.

Our future pebble atoms are somewhere among these sub-terranean rock-currents. The currents ran faster and hotter than they do today—perhaps about twice as fast. The deep Earth was hotter then, partly because its natural radioactivity was higher, and partly because of the residual heat generated by the impact with Theia. Evidence of this can be seen in some of the volcanic rocks that they generated: lavas termed komatiites, denser and richer in iron and magnesium than today's lavas and so betray-ing higher temperatures of formation. However, as far as can be judged from the oldest surviving remnants of crustal rock from that time, the basic pattern of plate tectonics operated in a fashion that was broadly similar to the pattern we see today.

The pebble atoms, then, were mostly hundreds or thousands of kilometres underground, scattered somewhere between the surface of the Earth and the surface of the core. There, they circulated, bound up in the minerals of the mantle rock, wait-ing some 3 billion years for their release. This mantle is mostly solid rock, but nevertheless it is rock that flows slowly when put under sufficient sustained pressure, as do the equally solid crystals of ice in a glacier today. The pressure exerted on a glacier is gravity, and the motion is downwards. The force exerted on mantle rock comes from heat generated by radio-activity, heat that increases downwards within the Earth. The heat—particularly at the mantle's contact with the much hotter core—makes the rock less dense, causing it to rise towards the surface. Then, when that slow fountain of rock has ascended high enough to cool and grow more dense, gravity takes over, and the rock-material begins to fall back towards the core under its own weight. One such cycle of ascent and descent takes hundreds of millions of years. A part of the mantle that is now close beneath the crust would have started its upward journey when dinosaurs walked the Earth.

There are faster routes within the solid Earth, superhighways of rock that are termed mantle plumes: hotter, faster fountains of upwelling rock several hundred kilometres high and one or two hundred kilometres across. One of these, that first reached the surface of the mantle some 60 million years ago, is still lifting the crust beneath Iceland (and Iceland itself, of course), causing it to be about 3 kilometres higher than it would otherwise be. Another plume lies beneath Mauna Loa on Hawaii. The extra heat it supplies has generated the lavas that, pouring to the surface, have made Mauna Loa, in a brief few million years, the largest volcano (and the largest moun-tain, measured from crater summit to its base on the sea floor) on Earth.

Since the time of the initial formation of the solid Earth, the pebble atoms have been bound up in minerals: specific min-erals that reflect molecular arrangements that not only suit the chemical affinities of the various elements—positively charged silicon and aluminium atoms allying with negatively charged oxygen atoms in silicate minerals, say—but that also that suit the conditions which prevail thousands of kilometres down in the Earth. These are ones of high temperatures (currently between 1000 and 4000°C, and in those far-off times, a little hotter) and also crushing pressures, equivalent to many thousands of atmospheres, causing the molecular frameworks for mantle minerals to be tightly packed.

We can manufacture tiny amounts of mantle-like minerals today, very briefly, in the laboratory, by mimicking the conditions present in the deep subterranean Earth. To do this, geologists put a pinch of surface minerals between the minute diamond jaws of a tiny (but very powerful) anvil, heat it to the appropriate temperatures and squeeze—hard. One can 'see' what happens by shining X-rays through the tiny mineral grains, and looking at the pattern that emerges—a pattern that reflects not the outer shape of the grains but the molecular framework within. To make the experiment more realistic, the grains have to be of the appropriate chemical composition for the Earth's mantle—enriched in iron and magnesium, with less silicon and aluminium.

New minerals emerge—often with a palpable 'pop' as the atoms collapse into new, more compact frameworks. The minerals at a thousand kilometres and more down have unfamiliar names, such as perovskite (and, deeper still, post-perovskite) and ringwoodite and ferropericlase. One or two are more familiar—carbon, for instance, forms the cage-like and highly sought-after molecular framework of diamond.

The pebble atoms were bound in these highly pressurized molecular structures for some 3 billion years, travelling at a few centimetres a year. We do not know quite what form the

26

mantle currents took. After all, it is impossible for us to travel physically into the mantle today, to see how it might behave. Our knowledge derives from such information as can be gleaned from the anvil-experiments, and from tracking the progress of earthquake waves through the Earth, and from seeing what kind of blocks of mantle rock are brought up sporadically by volcanoes. The earthquake-wave data clearly shows that the mantle today comprises two main shells—an inner and an outer mantle of different densities. The boundary between them probably represents a 'phase change', separating minerals adapted to the different pressure–temperature environments. Do the rock currents of the Earth's mantle today cross this boundary at will, or are there two separate systems, or are there two mostly distinct systems with leakage between the two? This point is still being debated.

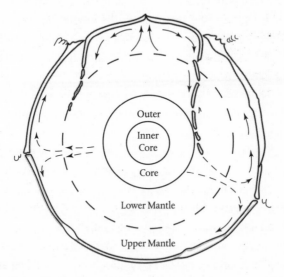

FIGURE 1 The Earth's structure, and currents that move within it.

If the Precambrian mantle functioned like this, the pebble atoms would have—on ascending or descending, every few hundred million years—readjusted, changed to a different mineral-structure, sometimes in a chain-reaction involving so many adjacent minerals that the resultant 'pop' became an earthquake tremor sufficiently strong to shake the Earth's surface.

At those depths, the temperatures are typically a few thousand degrees—easily sufficient, if at the Earth's surface, to melt all of those minerals. However, in the mantle, the minerals are mostly kept in solid state—just—by the immense pressures that are present at those depths. Indeed, on descending deeper into the Earth, the continuously increasing pressure just about keeps the mineral matter solid throughout the entire thickness of the mantle—with a few exceptions, as we shall see. Then, at the base of the mantle, comes the dramatic change at the core/mantle boundary (the temperature here being about 4000°C) where there is an abrupt change to a dense, liquid state (detectable at the surface because it blocks some types of earthquake waves) that is consistent with a composition of molten iron and nickel. There is a steep temperature jump here also, the silicate rocks of the mantle acting as a gigantic thermos flask around the metallic core.

In places within the mantle, too, the heat is just sufficient to overcome the mineral-stabilizing effects of pressure, the atoms vibrating fast enough to break their molecular bonds and to transform into a melt. But this is not true of all the minerals, just those whose bonds can be more easily broken—that is, that have a lower melting point. And so pools of underground magma can gather. These are generally less dense than the surrounding rock from which they had melted, with more

28

silicon and less iron and magnesium. They are set to rise to the surface, if they can find a route upwards.

ASCENT

The bulk of the future pebble atoms were gathering, ready for such an ascent, sometime between 1.5 and 1 billion years ago. The mantle currents were slowly moving them into a relatively small volume of mantle, likely wedge-shaped, perhaps several hundreds of kilometres long by several tens of kilometres in thickness—still dispersed, but occupying a smaller volume than at any time in their past. This would not have been quite all of the atoms of our future pebble, but the greater part of the rock-forming atoms would have been there: the silicon, and oxygen (mostly bound up with silicon as silicates) and the aluminium, and likely the iron and magnesium too.

Other atoms in our pebble, though, will have taken very different journeys, and set off to the Earth's surface at very different times from that of the main cohort. Chief among them are those destined to travel through the oceans: sodium and potassium and carbon and chlorine—and through the atmosphere—some of the carbon also. A very few atoms might, just, still arrive from space. But we will stay with the bulk of the pebble atoms. They are underground, waiting for their release to the surface.

The approximate billion-year date of this release is readable to geologists today by a very specific clock that depends on the peculiar behaviour of a pair of the lesser-known elements: neodymium and samarium, two of a series of closely related elements classed as 'rare earths'. These elements are so closely related that, where one goes, the other tends to go also—except

29

in that extreme environment where molten rock is generated in the mantle. Here, the relatively silica-rich melt takes with it rather more neodymium than samarium.

So far so good, but there's another twist that provides the mechanism for this particular atomic chronometer. Neodymium comes in several isotopic forms, each with their different numbers of neutrons to go with their constant 60 protons each. One of these, ^{143}Nd, has 143 (protons + neutrons), and is produced by the decay of one isotope of samarium, ^{147}Sm, that has 147 (protons + neutrons). The decay process happens *extremely* slowly, and is unaffected by changes in temperature, or pressure, or chemical circumstance. This is strictly the business of the nucleus, to which the state of the world outside is irrelevant. Now, ^{147}Sm has a half-life of 106 billion years, which means that after this interval of time half of any original amount of ^{147}Sm will have decayed into ^{143}Nd by the loss of an alpha particle (effectively a helium nucleus with two protons and two neutrons).

From then on, in the newly formed, relatively samarium-depleted melt (and in the piece of crust that it will become), less of the ^{143}Nd isotope is produced, and this particular isotope then forms a progressively lower proportion of the total neodymium in the rock. The timing of this slowdown of ^{143}Nd production can be deduced from very careful analysis of the contents of the various samarium and neodymium isotopes in a sample of crustal rock.

It is a very subtle atomic clock, but it is a generally reliable one, and it has come to be widely used in working out quite when any rock sample was ultimately extracted from the mantle. Remarkably, its workings (unlike those of most other atomic clocks) are but little affected by all of the subsequent history of that rock because of the way that the 'parent' element samarium sticks close to neodymium, its daughter element, during episodes of further melting or deformation or erosion. Thus, such a 'neodymium model age', as this particular clock is termed, can even be read—with a little caution, as we shall see later—from analysis of our slate pebble. It is a classic example of the levels of ingenuity (or low cunning) that earth scientists have had to acquire to sort out the really big questions in geology, such as how old any continent really is.

So—the magma-to-be-our-pebble is gathering somewhere in the mantle, far south of the equator (as its future travels across the globe will indicate), perhaps at the latitude of, say, Tierra del Fuego today. Its longitude then, by the way, is far harder (currently virtually impossible, indeed) to pin down. The surface onto which the magma is to ascend, in the Precambrian times of a little over a billion years ago, was one that is to us only moderately alien, at least by comparison with the dramatic events that had taken place around the original accretion of the Earth. A lot had happened in the 3.5 billion years of the underground sojourn of our pebble ingredients.

One billion years ago, there were continents and oceans—though in this case 'continent', singular, might better fit the bill, for this was one of those times when most of the Earth's landmasses had aggregated into one supercontinent, called Rodinia, surrounded by a mighty ocean, sometimes termed Mirovia. A good deal of Rodinia consisted of the landmasses (already ancient) that we now know as Africa, Canada, Greenland, Siberia, and Australia. But, we would scarce recognize them, for although their continental cores had been long formed, parts of them—the Rocky Mountains, the Urals—were still, like our pebble, features of the far geological future, their matter being still imprisoned deep within the Earth. And, the parts of them that were then essentially formed and that we now know as, say, the Canadian Shield, were then still deep underground, beneath towering mountain ranges, waiting to be revealed by a billion and more years of erosion, as those mountains were worn down to their roots.

In the atmosphere, oxygen was beginning to accumulate, but we as visitors (if a time machine could be invented) would likely find its levels insufficient—akin, perhaps, to levels at the summit of Everest, while carbon dioxide levels would have been several to many times higher. Not poisonous in the short term in itself, perhaps (unlike the case with carbon monoxide), but very uncomfortable. It would have been like being in a very stuffy room, or working all day in a brewery: one would be easily out of breath, while one's blood would become more acid.

This was a world in which life had been long developed—for 2 billion years at least—but it was microbial life. Multicellular organisms were still—perhaps because of the low oxygen levels—hundreds of millions of years in the future. So, no fish, no crustaceans, no worms, no trees or flowers or grass. Microbial life does not mean simple life. These Precambrian microbes aggregated into mat-like structures, which would then, as now, have been of great intricacy (that layer of green scum on the bottom of the rainwater puddle on your patio and the bacterial film on your teeth are phenomena of stunning multi-species complexity). Some of these microbes were likely huge—there is a gooseberry-sized microbe (a gromiid) today that rolls along the sea floor leaving a distinctive trackway that looks a little like a miniature bicycle wheel track (from a bicycle, though, with bald tyres). Similar trackways have been found in billion-year-old strata in Australia and India, but they used to be ascribed to the precocious development of worm-like organisms, before this giant rolling microbe was discovered making its tracks on modern sea floors.

THE NEW CONTINENT

This, then, was the world that lay above this volume of mantle rock that would release most, but not quite all, of the components of our slate pebble. These pebble atoms were an infinitesimally small part of a mass of rock material that was gathered to form a small, new continent, a minor addition to the Earth's crust but (arguably) a significant one. It was one on which—much later—King Arthur would reign, and Shakespeare would write sonnets, and a revolution would start that would spread factory chimneys and iron foundries across the world. That crustal fragment would, in its early days, stretch as far as Newfoundland—but not as far as Scotland. It is now called Avalonia.

What was it, then, the magic ingredient that tempted the matter of the pebble-to-be, and many trillions of tons of mantle material around it, to ascend to the surface, and to create the nucleus of the new continent Avalonia?

A continent is not so much a large piece of land that is above water, as a piece of crust—of whatever size—that is light enough to float above the denser rock of the ocean crust. Continents are not quite indestructible. The wind and the weather slowly wear at them. But, they are built to last much longer than does the ephemeral ocean-crust, which barely manages to exist for more than 100 million years (or 200 million, at the most) before it is forced back into the consuming mantle at ocean trenches.

Where does one make a continent? A good place to start is at the place where one tectonic plate plunges underneath another, at the ocean trench that is the surface expression of a 'subduction zone'. This is where a piece of old ocean plate is forced back into the mantle, to ultimately be recycled into solid Earth matter. The old ocean plate is cold—so one would think that it would make the adjacent mantle less likely to melt. However, it is also wet, carrying with it some ocean-water in rock fractures and in the spaces between sedimentary grains, and also chemically bound within hydrated minerals on the surface of the ocean crust. At depths of 50–100 kilometres, that water is released into the mantle that lies above the descending plate, and it is at this point that it catalyses the formation of a new continent.

The water released from the descending crust, dissolving into the mantle-material, has the effect of lowering the temperature at which this material begins to melt. This is the trigger that is to release the pebble atoms, and trillions of tons of their kin, from the solid scaffolding of the mantle minerals, into pools of underground magma. The new magma was silica-rich, water-rich, mobile, and less dense than the unmelted mantle rock (especially as the latter had been made more dense by having some of its lighter constituents removed from it). It sought, and found, conduits to ascend towards the surface.

A billion years ago, somewhere off the coast of the new supercontinent Rodinia, out in the ocean, a chain of new islands were forming. These were the beginnings of Avalonia, and they would probably have looked something like the Marianas Islands in the Pacific Ocean today, with Guam at its southern end, or like the Lesser Antilles in the Caribbean. Parallel to the chain and a hundred kilometres or so away (though invisible, of course, at the surface), is a deep ocean trench. This is where the ocean crust is descending.

The volcanoes on such a chain of islands are decidedly not the spectacular and photogenic, but relatively harmless, kind of lava flows that one sees in film footage of, say, Hawaii. Rather, they are hyper-violent: the viscous, silica-rich magma does not usually flow out, but is ripped into tiny shreds of ash by massive explosions. The ash is then either carried tens of kilometres into the sky as eruption columns by the enormous heat released, or speeds along the ground in fearsome pyroclastic density currents—dense hurricanes of incandescent ash capable of instantly sterilizing an entire island—as opaque ash clouds above plunge most of the landscape into pitch darkness. Periodically these growing islands are shaken by powerful earthquakes, and swept by tsunamis, the lethal but inescapable phenomena associated with the growth of what is termed an island arc, a factory for the production of new continental crust.

Ocean crust can descend also beneath a continent, and the continent-forming (or rather, in such a case, continent-enhancing) volcano-bound magmas simply punch up through the crust of the existing continental mass. This is how the volcanoes of the Andes, for instance—Cotopaxi, Chimborazo and so on—are growing on (and adding to the mass of) the western edge of the South American continent. However, in such a case the newly ascending magmas mix with, and are contaminated by, the much older continental crust that they traverse. This seems not to have been the case with the Avalonia: such contamination would have been revealed by those remarkable neodymium model ages, which show a pattern sufficiently regular to betray the original Avalonia as pure, uncontaminated crust. These, then, seem to have grown in isolation as islands over an oceanic crust, and it is here that our pebble material made its entrance at the world's surface.

Most of the atoms of our future pebble were in that billion-year-old island chain somewhere. Some of them will have been erupted as ash particles, or as crystals in the short stubby flows of silica-rich lava. Some have not made it yet to the surface, but are trapped underground still, a few kilometres below the surface, in chambers where magma ascended only part way, and then cooled and solidified in place. The particles of which our pebble will eventually be constructed were still widely scattered, likely among several islands of the chain; they are now slowly nearing each other, but there is still a long and tortuous path ahead before they come together to occupy the same volume as has, say, this book in your hands.

These volcanic rocks of the original continental nucleus were made of various crystals, together with some natural volcanic glass where the magma was quenched by cold air or water before crystals had a chance to grow. Yet few of these rocks, or of their component crystals, were to survive in anything like their original form to be recognizable by geologists, 1 billion years later, as parts of the primordial Avalonia continent. For these rocks were still to undergo much history, and be thoroughly refashioned, before they released the pebble particles into what one might call their decisive journey, if not their final one. For the travels—and travails—of Avalonia have scarcely begun.

Distant lands

It was Avalonia, but it could as well have been Shangri-La, or Conan Doyle's *Lost World*. The vanished nursery-ground of our pebble—the land-that-once-was—is now terribly distant from us. Epic journeys are needed to reach it, back into deep Earth time. It is not a physical journey, as such, to be taken using helicopter and dug-out canoe, by explorers in pith helmets, wielding machetes. Rather, it is a journey of the imagination, albeit one grounded in the physical reality of this lost continent, and the traces of it that remain.

Touch the pebble, and you are touching Avalonia, in the form of the mineral grains that represent its destruction, a continuous dismantling accomplished by half a billion years of wind and rain and flood. It is through those grains that one needs to search back for the landscape that they once represented. Or rather, *landscapes*. Avalonia was not a single unchanging entity, which we can hope to picture in ever-more-faithfully captured detail as we study its ancient past. This lost continent ceaselessly changed, mutated, renewed itself. And the tiny mineral fragments that now form part of the pebble are not so much fragments of *it*, as fragments of *them*, of its many changing faces. One might, for comparison, take tiny relics from each of the seven successive cities of Troy (and sample, too, the modern buildings that now stand above their buried remains). Grind, then, those fragments into a fine powder. Then, give a handful of this dust to an archaeologist, and say 'now, bring those cities back to life!'

Landscapes are transient. This is a concept that does not come easily to us. In our brief lifetimes we see the Earth's landmasses as things of massive permanence, the bedrock of passing civilizations. And yet even in these human lifetimes we can see masses of rock debris piled up beneath mountain crags—and, as we walk nearby, hear the fall of new scree fragments, dislodged from rock faces by wind and water. We see sand move along a river floor, driven by the flowing water. Occasionally, we might see villages smashed by floodwaters, the wreckage littered with mud and boulders carried from miles up the valley. Perhaps we may see volcanoes erupt, to bury landscapes under ash or lava.

Multiply such changes by the vastness of geological time, and there is plenty of time to change the face of a continent. This was something that one of Charles Darwin's colleagues (and mentors, indeed), Charles Lyell, realized, when he coined that rather inelegant term 'uniformitarianism' to convey the idea that slow everyday processes can, over the passage of aeons, utterly transform the face of a planet. For this, he needed the concept of earthly aeons (rather than the few thousand years of a Biblical timescale). Here, Lyell was most clearly indebted to James Hutton, an 18th century savant, who, seeing ancient rock strata of the Scottish mountains lying on top of the eroded roots of a yet older mountain chain, suddenly realized the immensity of time ('without the vestige of a beginning, or prospect of an end') represented by that simple physical juxtaposition of rock layers.

They were characters, those Victorian scholars, whose spirit you might not glean from the stern portraits of that era. Lyell, for instance, when learning of Darwin's theories of how coral atolls formed, 'was so overcome with delight that he danced about, and threw himself into the wildest contortions, as was his manner when excessively pleased'. Now *that* sounds like a man that it would be a pleasure to share a beer with, after a long day gathering fossils in the hills.

Lyell would have danced all night, for sure, if he could even have an inkling of just how much of the lost world of the past is preserved in its detritus, encapsulated in the pebble of our story. He might, though, have become a touch rueful, and slowed his tempo to a stately and pensive gavotte, if he knew that his cherished concept of 'uniformitarianism' would ultimately show its limitations, brilliant and influential though it was. The lost worlds of Earth were often so different from our current world that the present is not always a reliable key to the past. The pebble holds strange worlds within it.

QUARTZ TALES

What then do we see in our pebble, of those shards of a vanished Avalonia?

One can start by holding the pebble to the light and peering at it through a simple magnifying glass. It is best to wet the pebble first, in river or sea, for this makes the mineral structure of the pebble more easily visible. Focus on the paler, slightly rougher-textured stripes. They will show tiny rounded outlines packed together, each a fraction of a millimetre across. These were once silt and sand grains, washed out of Avalonia. The pebble contains many thousands of them.

To get a closer look at them, we need to enter the Lilliputian world of the mineral kingdom. You can start by looking at the pebble surface with a standard optical microscope. This helps a little, but not as much as you might think. The tiny translucent, grains, squashed together, reflect light from a myriad grain boundaries within the outer part of the pebble. The overall effect is blurred and hazy—one cannot see individual trees for the forest. A century and a half ago, Henry Clifton Sorby, a Sheffield scientist, pioneered a better way. He cut rock fragments in half, stuck one half, cut side down, on a glass slide, and then carefully ground away the rock until it was *almost* gone. When the rock was down to a wafer only a thousandth of an inch thick, it became translucent. The grain shapes, now in outline, became beautifully visible under the microscope. For its day, it was a radical innovation. Too radical, perhaps, for Sorby was initially ridiculed for 'studying mountains with a microscope'. Sorby had the last laugh, though: such manufacture of 'thin sections' of rocks remains a standard technique even today.

A thin section of the pebble can now show, very nicely, the size and shape of these tiny grains. It can tell you of what minerals they were made, by the way the rays of light coming through the thin section interact with the molecular structures that they pass through. Here geologists do not use an ordinary optical microscope, of the kind you can find in a biology laboratory. They use one in which special filters polarize the light—that is, they force the light to vibrate in only one direction. The polarized light, after its travels through the thin mineral slices, takes on various—often strikingly beautiful—combinations of colours and shades. Each such combination, to a skilled geologist, betrays a specific type of mineral.

Most of the mineral grains of the pebble, examined this way, are glass-clear, and then, when the light passes through not one but two polarizing filters, one above and one below the mineral slice, they appear in various shades of grey. This is quartz, or silicon dioxide, the mineral that typically makes up the greater part of most natural sands. You probably carry around a quartz crystal with you, carefully sculpted to vibrate at 32,000 times a second when a small electric current is applied to it. This vibration is then translated, by the commonplace miracle of microelectronic circuitry, into the exact time of day on your wristwatch.

Quartz is nothing if not versatile, for it can be found in rocks that have crystallized from a magma (that is, in igneous rocks), and in those that have been entirely restructured by heat and pressure at depth (the metamorphic rocks) as well as in sedimentary rocks. It can form lovely crystals, clusters of hexagonal prisms with elegantly pointed ends, when it crystallizes from hot water within rock cavities and mineral veins deep underground. These crystals can be colourless, or if they include tiny amounts of different impurities they can turn purple (in the form known as amethyst), or rose-coloured, or smoky brown, or indeed—bewilderingly for a novice mineralogist—any colour of the rainbow—and quite a few more. (In microscopic thin section, though, these colorations generally disappear, or remain as only the slightest of tinges.)

Most quartz, though—say that which crystallizes from the kind of silica-rich magma that cools to form a granite rock— forms rather shapeless grains, nothing like those that might adorn a mineralogist's cabinet. Quartz is, in truth, mostly a rather self-effacing mineral. The neighbouring crystals in a granite—of mica and feldspar—grow earlier in the cooling magma, and express their crystal shape well. The latecomer quartz fits in around those earlier crystals, and moulds itself to their shapes. Internally the quartz still remains crystalline, of course, with its own precise molecular scaffolding of silicon and oxygen atoms.

It is a late arrival, perhaps, but thereafter quartz long outstays its neighbouring minerals. For those other minerals, born at higher temperatures, do not easily tolerate the cold and damp of the Earth's surface. There, the molecular structure of these minerals is put under strain and—when there is water around to speed the process—they snap, and the minerals disintegrate. The disintegration is complex, and fruitful, and fascinating. But it has little effect on the quartz. As the rest of the rock crumbles around it, this durable mineral is released as grains into soil or scree or river bed, to begin their long journey to the tiny patch of sea floor that is a pebble-to-be.

Quartz is a relatively simple mineral, yet each grain has its own character. Each can transport the enquirer into a different part of Avalonia. One quartz grain, say, might show a telltale waviness of its shades-of-grey pattern, when the mineral slice is viewed through polarized light. That grain was originally a quartz crystal caught up in a growing mountain chain, its molecular structure distorted by the immense stresses exerted there. Move to another quartz grain, one that has a distinctive mosaic pattern: here one is taken even deeper into the roots of the mountain belt—the tectonic shearing here was so intense that the quartz crystal snapped into a mosaic of smaller crystal domains. Some quartz grains are distinctive in having yet tinier crystals of other mineral within them. A common fellow-traveller is rutile (titanium oxide), seen as hair-like crystals sometimes so thin that they need the highest powers of the optical microscope to even make them visible at all. Yet other quartz grains are made up of clusters of smaller grains—that is, they are in reality tiny fragments of a more ancient sedimentary rock, originally derived from some igneous or metamorphic rock in some yet more distant time, and then recycled in turn.

It is like being in a room full of explorers, each of whom has a different story to tell—some of adventures in tropical jungles, others in polar wastes, yet others who piloted bathyscaphes to the bottom of the ocean. So it is with these humble quartz grains, born in many and various parts of Avalonia and, after long travels, now assembled in the pebble. There are other travellers there, in that pebble, too, that have even more exotic stories to tell.

TALES OF RARE MINERALS

Hidden among the abundant quartz grains, there lurk other mineral grains, less common, more distinctive. Each again has its own narrative. Often, the stories of the rare minerals are so eloquent that geologists go to unusual lengths to decipher them. Zircon is likely there, and tourmaline, topaz, rutile, garnet, staurolite, monazite, apatite—there are many types of these rarities. These minerals are normally so outnumbered by the quartz that, under the microscope, in a typical thin section, one might see only one or two grains, by chance. To find more, one needs to resort to more drastic measures. One has to crush the entire pebble into sand-sized particles, and then to drop these into a beaker of a heavy liquid, such as bromoform ($CHBr_3$). In this high-density liquid, the quartz and the other minerals simply float to the top. The rare minerals we seek, though, have such high densities that they sink through to the bottom of the beaker, and can be retrieved (and hence they are often called 'heavy minerals').

In this way, one can amass enough of these rarities to make an analysis. Their identification is typically done using a simple binocular microscope: it is an old-fashioned procedure, and one that needs much skill and experience to distinguish subtleties of colour and lustre and surface texture among the menageries of different minerals that might be found here.

Or nowadays one can substitute classical mineralogical skill with technology—albeit a technology that requires its own expertise. Take those grains and put them into a scanning electron microscope, or into its cousin, the electron microprobe, both machines that are now standard workhorses of the geological laboratory. With either of these, one can fire a tightly focused beam of electrons at each grain, to cause it to emit characteristic patterns of radiation that betray the types of atom within it, and hence allow the mineral to be identified: zirconium in zircon, for instance, and lanthanum and cerium in monazite, titanium in rutile, a combination of calcium and phosphorus for apatite, and so on.

By whichever means, one now has yet another set of clues to the vanished landscapes of Avalonia, an infinitesimal part of which now lies compressed within the pebble. For zircon crystallizes within granite magmas and within the hottest parts of the cores of mountain belts, where the rock almost (but does not quite) melt. Garnet is also typical of metamorphic rocks, but not of such extreme conditions—a mere simmering at some 500°C and at depths of 10–15 kilometres will do. Apatite (calcium phosphate) can—and must—crystallize in our bones to give them strength, but as a heavy mineral it typically formed in granites at temperatures reaching 900°C.

With some of these minerals, one can build some quite specific histories. Zircon is unusually eloquent in this respect. It is zirconium silicate—that is, a mineral combination of the elements zirconium, silicon and the near-ubiquitous oxygen. But it also shelters other elements, elements that otherwise find it hard to find a mineral home. These are large ungainly atoms such as hafnium, yttrium, thorium and uranium. None of the main rock-building minerals like to accommodate these elements—not olivine or pyroxene, not mica, or feldspar, or quartz. They are atomic wallflowers, so to speak, and only find a compatible mineral partner when zircon starts to crystallize.

They repay this hospitality by providing a picture of the inner workings of a magma chamber, as revealed by the near-supernatural capabilities of the scanning electron microscope. If a zircon crystal is polished and electron beams are fired at it at an angle, some will bounce off the polished surface, to be caught by a specially placed detector. The denser the mineral (in reality, the more atoms there are with large atomic nuclei), the more electrons will bounce off, and so the image captured by the detector will become brighter. Many zircons, viewed in this way, show remarkable and beautiful patterns of concentric bright and dark rings. The brighter layers typically have more hafnium, while the darker ones are poorer in this element. These provide a kind of tape recorder of the growth of each crystal, as the magma continually changed composition

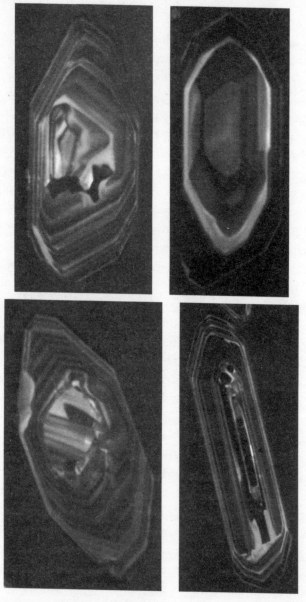

FIGURE 2 Zircon crystals, as imaged by a scanning electron microscope; each has a different history, as revealed by its pattern of growth lines.

around it. It is like tracking minute fluctuations in weather—only in this case the weather patterns are of a long-destroyed magma chamber, which was once several kilometres below the surface of the Earth.

But this fragment of history is, to some extent, a sideshow. Zircon's main contribution is to allow students of the Earth—or of a pebble—to navigate, with incomparable precision, through the fourth dimension, through the deep time of Earth history. It does this courtesy of a transmutation, a kind of natural alchemy that the medieval scholars never dreamed of—even as it was continually happening under their noses (albeit in the realm of the very small), while those ancestors of today's scientists made ever more frantic and fruitless attempts to conjure its action to make alchemical gold. The key here is zircon's hospitality to uranium.

THE TIME MACHINE

Uranium, as Henri Bequerel accidentally discovered in 1896, is unstable, or 'spontaneously radioactive' as he termed it when he discovered the mysterious radiation that could travel through thick layers of paper and yet still fog photographic plates that had been entirely protected from sunlight. The largest of the common elements, uranium, has a variety of forms—isotopes—in which different numbers of neutrons struggle to keep the 92 protons in each atomic nucleus from flying apart. Ultimately, they don't manage it, and the nucleus shatters, with smaller or larger parts of it flying off.

With the commonest uranium, ^{238}U, which has 146 neutrons shepherding the protons, half of an amount of uranium has gone, transmuted into lead, in 4.46 billion years. For the kind that goes into atomic bombs, ^{235}U, with 143 neutrons, the rate is faster and therefore the half-life is shorter, at 704 million years (and hence this isotope is more radioactive, and more dangerous to us).

The lead that is the ultimate end result of this breakdown typically stays within the tough zircon crystal (a little uncomfortably, for zircon will not take up any lead—thankfully for its use as a time machine—while it is growing). Therefore, measure the amount of the different uranium and lead isotopes in each crystal using that modern atom-counting machine, a mass spectrometer; establish the half-life, as precisely as possible, of each isotope; and from these one can calculate back, in years, to the time when the zircon crystal was formed.

Cross-checks need to be made, of course, because some zircons, in their long sojourn underground, may have lost or gained tiny amounts of lead or uranium, to make the atomic clock tell time inaccurately. One can, say, compare the dates given from two isotopes of uranium within one zircon crystal—if they give the same result, the age is said to be concordant, and hence likely to be trustworthy. If they differ, giving a discordant date, then something has upset the workings of the atomic clock, and that result may be discarded.

Even with all the care and caveats that are needed, this is a precious gift to a geologist. For most minerals are mute as regards time. All those quartz grains may tell their various stories of cooling magmas or crystallizing mineral veins, but they do not tell us *when* these events happened. The quartz in a grain may have crystallized a million years ago or a billion years ago, and we simply have no means of telling. It is a frustrating inability when one is trying to reconstruct the history of an eventful planet (though quartz, in fact, does keep a certain kind of Earth time, a *much* more recent one, as we shall see further in our narrative). Minerals such as zircon open the doors to a time machine, one that can traverse an entire planet from its beginnings.

Today, the dating of zircon crystals has become, for geologists, the main chronometer for events in deep time, and a 'good' age will be accurate to quite a bit better than one per cent of the total age, that is to within plus or minus a million years (or less) for crystallization events that happened hundreds—or thousands—of millions of years ago. Immense efforts are made to squeeze ever-so-slightly better levels of precision out of the zircons: to design better atom-counting machines, and to better establish the radioactive decay rates of the different isotopes. Or one can, very gently, remove the outer parts of zircon crystals, where contamination is most likely. This is a lot of effort—but time is important, after all. One cannot be too careful with it.

Our pebble will probably contain something between a few dozen and a few hundred individual zircon crystals. Each can be individually dated. Each date marks, precisely (to that one per cent-or-less) the growth of a single zircon crystal, in some magma chamber or in the roots of a growing mountain chain, both now long-vanished from the Earth. But how big a magma chamber, or how long a mountain chain, was that? Here geologists play the numbers game. One dates many zircon crystals in a sample, and sees what kind of range of ages is obtained, and how these are clustered. These will give a view, as through thick mist, of huge and important events a long time ago.

Our pebble would probably yield what has come to be known as a typical Avalonian signature. Quite a lot of zircons would be a little over 600 million years old, and some will be 700 million years old and more. These are usually thought of as the result of phases of huge volcanic outbursts on that continent at those times, linked to solidifying magma chambers at depth, each phase lasting tens of millions of years. Then, there is a scarcity of such evidence until a billion years ago, when another cluster of zircons appears. It might be that a few very old zircons turn up—even older than the continent itself at, say, two billion years. How can that be? Well, Avalonia, on its travels, nudged against or joined with more ancient pieces of crust (such as what are now parts of Africa), from which some sediment (including those very ancient zircons) was then swept. Equally, populations of Avalonian zircons can turn up on the remains of other former continents, as a kind of mineral calling card.

This kind of history is smeared out to the extent that individual volcanic eruptions count for little, even were they to be as big as Krakatoa. It is the broad sweep that counts, within which details are lost. Even more lost is the shape (or, of course, *shapes*) of that evolving Avalonian landscape, which yielded those zircons. When the landscape has been ground up so thoroughly that only fragments of less than a millimetre are left (albeit that there are now trillions of them), then what hope of saying whether those volcanoes were beautifully symmetrical like Fuji, or partially collapsed, like Mount St Helens? Tiny fragments of Avalonia survive, and some survive beautifully to tell long and intricate tales. But the successive surface contours of those landscapes have gone, irretrievably. Such is the fate of any landscape raised above the sea, into the mill of erosion.

THE BIRTH OF MUD

Another ingredient of the pebble results from that destruction. It is a major one—the grey stuff of this grey rock. It far outweighs the stuff of the pale sand-rich layers. But it is one that is surprisingly difficult to reconstitute back to its original form. It is mud.

For the pebble is no longer muddy. It is a slate. The original sediment has been squeezed and heated into hard rock, and the material that was once mud has changed its nature. In thin section, under the microscope, the shades-of-grey of the quartz grains are outnumbered by countless blade-like crystals of brilliant colours—reds, greens, yellows, blues—when placed between the polarizing filters. These are now, essentially, types of mica crystal. But they were once much, much smaller: the submicroscopic clay particles of muddy Avalonian soils and sediments.

When wind and rain wore down the landscapes of Avalonia, the attack was as much chemical as mechanical. The rain was acid, even more so than today's, for the atmosphere half a billion years ago was—as far as we can judge—richer in carbon dioxide than now (though our industrial civilization is catching up with impressive speed in this respect). Carbon dioxide dissolves in rainwater, to make carbonic acid. Just here and there, the chemistry of the water at the surface may have been altered too, by the beginnings of plant life on land. Today, humic acids released from live and dead plants at the land surface are corrosive towards many minerals. Then, there were no trees, no ferns, no flowers, no grasses. But, there may have been, in moist sheltered places, growths of single-celled algae, and fungi, and perhaps simple leaf-like forms representing today's liverworts. Even these pioneers among land plants could hasten the pace of mineral breakdown.

At the surface, the rock would corrode, turn crumbly as biscuit. This outward sign of mineral corruption does scant justice to the enormity of the change at the molecular level, where the cathedral-like vaults of the high-temperature minerals are dismantled to form flakes and sheets of the clay minerals, so tiny that a single gram of clay could be spread out (if one had sufficient patience) to cover hundreds of square metres. The tiny flakes, like their parent minerals, are essentially silicates or aluminosilicates: that is, their basic building blocks are pyramid-shaped clusters of silicon, aluminium and oxygen, arranged in sheets, on which are more or less loosely attached a variety of charged ions: calcium, magnesium and iron, potassium and sodium, and others besides.

Once formed, these flakes can be dislodged from the parent rock, and form part of a soil layer. The Avalonian soils were probably thinner than today's: with no forests and no grasslands, they probably had little humus. With no vegetation cover, they would move more quickly down slopes, by the action of gravity and by rainwater. Washed from that soil, they could be carried by stream and river water; dried, they could be wind-blown across large distances. The smallest particles are so light that they could traverse oceans to land on other continents, just as Saharan dust today can land on Europe.

The delicacy and lightness of these particles means that their paths soon diverge from those of the sand grains, co-liberated from the decaying rock faces. Once swept into a river, the sand grains will be driven by the current along the bed of the river, rolling and bouncing, typically travelling as ripples and dunes that migrate across river beds in a fast-flowing current. When the current slows, they stop. The clay particles, though, are carried in suspension above these stop-go sand carpets, travelling farther and faster, even moving in a slow current.

Only when the water stops moving, can the tiny clay flakes begin to settle.

In this way the components of what was once a single rock are separated during transport by flowing wind and water, sorted according to size, shape, and density, becoming more widely separated the longer the travel. In the pebble, the mud will have travelled on different paths from the sand grains, probably longer paths, and will have come from different parts of the Avalonia. A few particles may even have come from other continents, carried by winds high in the atmosphere.

There is another component of the pebble that will have a wider provenance, literally from any part of the Earth, for clay minerals are not the only product of the chemical weathering of rocks at the surface. As the original minerals break down, some of the atoms are released directly into solution as charged ions, particularly of sodium, potassium, calcium, and magnesium.

Once in solution, much of this material simply keeps on travelling down to the sea, in what we call 'fresh water'—that is nevertheless a complex chemical cocktail of dissolved salts, as one can see by looking at the label of any bottle of mineral water. These dissolved salts are what, over many millions of years, has made the oceans salty. Some of this material will find itself in the pebble too—but in this case a lot of it will *not* be from Avalonia, but from virtually anywhere on the Earth's surface. This is the most cosmopolitan component of any rock, and many global circumnavigations may have taken place before they eventually came to rest. Once in the oceans, the dissolved material can remain there, if not quite forever, at least for a very long time, and travel many times around the oceans, borne by ocean currents, before they come to rest.

How long dissolved material stays in the oceans, and how far it travels, depends on what the material is. Oceanographers studying this phenomenon refer to the 'residence time' of an element, which is the average time a particle of it spends in the sea. For some elements it is enormous. The average sodium ion for instance can travel the oceans for nearly 70 million years before coming to rest, simply because, while the oceans contain enormous amounts of sodium (about 10,000 million million tons, at the last count), they can hold yet more. With other, less soluble elements it is much shorter. The average aluminium ion spends only 200 years in the sea before dropping out of solution. For some of this material to find itself into the pebble, a carrier is needed: a dissolved magnesium ion, say, can attach itself to a passing clay flake. For others, a far more complex process is needed, the one we call Life. But that is part of a future story.

So here we leave this section. Somewhere on the northern margin of Avalonia, rivers are entering the sea. Those rivers are carrying countless grains of sand and mud into that sea—a sea that has been filling with dissolved salts for some 3 billion years. And from now the sea must carry those particles to their final resting place.

To the rendezvous

FORCES OF NATURE

Before any great expedition, there is a gathering of all of the forces—of the clans, the troops, the mercenaries—from near and far, by various routes. Once met, they will then travel *en masse*, their fortunes from then to be bound together, for good or ill.

Sediment particles of the future pebble were gathering, around the shores of Avalonia, in the Silurian Period, for a journey that would take them to a resting place, one where they would not see the light of day for something over 400 million years. The grains of sand and flakes of mud, with all their variety and histories, were being washed into some long-vanished shoreline by Avalonian rivers, rivers that have not yet been discovered, or charted, or named, by modern-day explorer–geologists. Likely these rivers never will be charted, for in flowing they eroded themselves away, washing away their own tracks, as Avalonia was being dismantled, grain by grain, by the eternal, tireless action of the weather. All that is left is the freight they carried, the baggage of sand, mud and pebbles.

The ancient shoreline lay not much more than 50 miles away from what is now our pebble beach in west Wales. It lay to the south, around what is now Pembrokeshire in South Wales. What did it look like, that ancient coastline? Well, it may even have resembled the rugged Pembrokeshire coastline of today, though it faced north rather than south, looking across an area of open sea that was later transformed into the Welsh mountains.

For the pebble stuff, the passage across that coastline marked the entrance into a new realm. As the river waters entered the sea, their onrush slowed. The sediment grains, no longer driven by river flow, would have piled up around river mouths as deltas, or within silting-up estuaries. They would not have been stilled for long though, for coastlines are places where energy is exchanged. New forces acted on these sediment particles: wind and tides and waves, the forces that nowadays mariners need to respect, and understand, and predict. Had any mariners been suddenly time-transported, boat and all, on to a Silurian coastline, the winds would have functioned much like today. But Silurian tides may have caused a discerning mariner to raise an eyebrow. For they were a little more forceful, then, than tides are today.

Tides, mind, have been perplexing philosophers and savants, and (in a more practical sense) seafaring people for much of recorded history. Just why does the level of the sea rise and fall with such regularity, to flood the land and then to ebb back, creating currents as they do that can first drag you inland, and then sweep you out to sea? Myth and legend saw this phenomenon as due to the breathing, or perhaps the heartbeat, of an enormous sea god. But our human ancestors also paid great attention to the sky—a sky that was large and bright and pervaded human consciousness in a way that is hard to imagine in our urban world of omnipresent artificial light.

It was clear that the pattern of tides had something to do with the phases of the moon: Pliny the Elder, for example, recorded the clear relationship between the two, as did his near-contemporary Seleucus, who thought that the Moon pressed down on the Earth's atmosphere. The Venerable Bede, a scientist as well as theologian, observed that the passing high tide was a huge, progressive wave that moved along the coast of eastern England.

It was a French savant, though, the Marquis de Laplace, who essentially solved what he called the 'spiniest problem in all of celestial mechanics'. He recognized that the tide was a bulge of water drawn up not only by the gravitational attraction of the Moon on the waters of the Earth, but also by the centrifugal force that prevents the two whirling bodies from falling into each other—and hence there is a bulge of water not only on the side of the Earth nearest the Moon, but also on the opposite side as well (to give two high tides and two low tides every day). The Sun has an effect too, which can make tides higher (as spring tides) or lower (as neap tides) depending on whether it adds to the lunar tide (when it is line with the Earth–Moon system) or diminishes it (when the Earth, Moon, and Sun form a right angle).

And so, as these vast bulges are raised in the Earth's watery envelope and travel across its surface, currents are generated, flowing towards the peak of the bulge as the water piles up in a high tide at any one place, and then ebbing away as the crest of the tidal bulge moves on. These are the currents that can overwhelm unwary travellers as they attempt to cross an estuary, and that sweep sediment particles back and forth, either taking them out to sea or piling them up near the coast as tidal sand and mudflats. The good Marquis was clever enough to work out the details of this complex machine of the tides at the Earth's surface, and also wise enough to leave Paris as the French Revolution was at its height—and so kept his head while so many others lost theirs.

It is almost a perpetual motion machine—though of course in this world one does not get something for nothing. And the world, hence, is slowing down. The energy that raises the tides comes from the momentum of the Earth–Moon–Sun system, and as this energy is used, the Earth spins more slowly, and the Moon moves farther away. So, 430 million years ago, the Moon was nearer, and days were shorter, with a little over 400 in a year. Tides were higher, and tidal currents therefore swept sediment more energetically across shallow sea floors than they do today

And then, these coastal waters in which our pebble particles find themselves were also agitated, more or less continuously, by the wind. That has been another long-standing mystery for those humans who sought to understand how the Earth functioned. What made the winds blow—sometimes gently (or not at all, on those calm days when the sea is like glass) and sometimes violently enough to snap the trunks of fully grown trees, or to dash a frail boat against a rocky coastline? And why did the winds come from some directions more often than from others? The crude maps of the ancients sometimes really did show a god or a cherub with cheeks ballooned out, blowing lustily across the Earth's surface.

Lacking the clear association that tides had with the moon's phases, and hence of a regular—and hence scientifically tractable—phenomenon, the winds were a real puzzle to our distant ancestors, and so a variety of gods were invoked as explanation, or at least as someone to blame. Njord in Norse mythology, for instance, or Vaju, the Hindu god of the winds. Stribog the Slavic god, was grandfather of the eight directions of winds. Fujin, in ancient Japan, strode around the world while carrying a sack full of winds on his back. A bewildering plethora of wind gods reigned in ancient Greece. There was Aeolus the storm god, son of Hippotes (not to be confused with Aeolus the son of Hellen, or Aeolus the son of Poseidon), who was keeper of the four winds, Boreas, Notus, Eurus and Zephyrus—north, south, east, and west respectively, that may have been winged horsemen, or horses—or, purely and simply, just winds. It was the gentle Zephyrus, for instance, given by Aeolus to Odysseus in a bag, who blew that hero eastwards towards Ithaica. Alas, the bag contained the other winds too. These additional winds, released by Odysseus's men just before their goal was reached, then blew them right back again. Aeolus was not amused.

As to more rationalist explanations—well, Xenophanes in the fifth century BC hazarded a guess that winds, together with rain, and streams and rivers would be impossible without 'the great sea'. Not strictly true (of winds, at least), but at least a stab in the right direction. Xenophanes, incidentally, had no truck with the idea of the gods as human in shape (would a thinking ox, he said, imagine an ox-god?), pondered on what fossils might mean (a world once covered by water, he thought), and in general considered that there was a truth of reality out there somewhere but that humans (yet) didn't have the means to discover it. A splendid man, from the sound of it.

It was not until the 17th century, in Italy, that Evangelista Torricelli put his finger on the true cause. Torricelli, a bright lad from a poor family, had a Jesuit upbringing in Faenza—or perhaps in Rome (the historical record is unclear). He grew to be an inspired mathematician, and became an admirer and briefly a pupil of Galileo. Best known for his invention of the barometer (and hence units of pressure are named *torr* after him) he may be regarded as the discoverer of nothing: he was the first man to create a vacuum—something that could be done, he said, despite Nature's usual abhorrence of such a state. Torricelli realized that winds are produced because different masses of air on the Earth are of different temperatures, and therefore at different densities—and hence these masses will flow relative to each other.

It is the sun (not the ocean, as Xenophanes thought) that drives this engine, heating the air in equatorial regions, so it moves upwards and then outwards, spreading out and cooling and sinking, being deflected by the Earth's spin, and spiralling into the patterns of wind and rain that we feel as the weather, day after day.

The wind, driving across the sea, catches the surface of the water, ruffling the surface into small ripples. These become larger as the wind continues to blow, and grow into small waves, and then into large waves, which move out across the surface of the ocean. Waves can travel for thousands of miles, before they break on some distant shoreline. The water itself does not travel such distances—it is a transfer of energy, not mass. As waves pass, the water simply moves in a circular orbit, about as big as the distance from wave crest to trough: one can see this by watching the movement of a piece of flotsam on the sea surface. In shallow water, these small-scale water movements can 'catch' on the sea floor, and move sediment particles to and fro. Sand is piled up into rows of ripples, while the flakes of mud become suspended in the water; they do not settle until they reach a sea floor not stirred by this action, either a sheltered lagoon or estuary, or the calm of a deeper sea, beyond the reach of wave-stirring.

Our pebble particles, in their transit from land to their resting place in the deep sea, would have been swept by the ebb and flow of the tides, and stirred by the waves, along that Silurian coastline, some 420 million years ago—gently on fine sunny days, and vigorously as ancient storms raged. On passing through these interlacing webs of shifting energy patterns, pebbles were separated ever further from sand grains, which in turn parted company from the delicate mud flakes. The various sediment particles, each reacting in their own way to the tug of the water, traced different paths along the sea floor—then joined together again in different combinations from different sources, at the whim of a rip current here or a tidal stream there. Jostling, bouncing, they would also change shape, become smoothed and rounded—particularly if they found themselves on the attritional, mobile conveyor belt of a beach.

And they would encounter life—a life that was there in abundance, in those shallow Silurian seas. The moving grains would, for instance, be pushed aside by the many pairs of legs that propelled the charismatic, extinct trilobites, that looked like so many overgrown woodlice, as they pushed themselves across the sea floor. Those trilobites were predators, near the top of the food chain. They were on the hunt for creatures that had an even more intimate relationship with the grains of the pebble-to-be: they ate them. For muddy sea floor sediment is nutritious—not so much in itself, but because it contains protein, fats, and carbohydrates in the remains of dead animals and plants, and also in the form of the microbes that are continually, busily dismantling those carcasses. This nutritious mix, a sedimentary broth, is eaten, digested and excreted by an army of worms and burrowing molluscs. As our future pebble particles travelled through the guts of these creatures, their composition was subtly altered: clay particles, in particular, have been seen to alter their chemistry, to exchange one set of metallic atoms for another, in the guts of modern marine worms; Silurian worms almost certainly had similarly transformational innards.

The interaction of our pebble particles with the strange life of those times was not always so passive, or benign. Sediment can be a killer as well as a life-provider. Some sea floor creatures do not eat mud, but instead shun it. These are the filter-feeders, which extend delicate tentacles to filter microscopic organisms out of the seawater. Load that water with sediment, from a storm or a river-flood, and the delicate feeding machinery is clogged and overwhelmed: the animals—corals, say, or brachiopods (lampshells)—suffocate or starve to death. It is a normal hazard of marine life, and is commonly graphically seen in strata, where fossilized colonies of filter-feeding organisms rest beneath the sediment layer that killed them. Among those grains in the pebble will be some—many, even—with just such a murky past: the killer grains, though, will look as innocent as their neighbours.

And the killer events were probably those that also moved the pebble particles closer to their final resting-place. A major storm is not just a bringer of mayhem to land and sea: it is also a mechanism to transport sediment yet farther out to sea. At the height of a storm, the winds may pile up a mass of water against the land, a mass that is made higher by the low pressure in the air above, literally sucking the sea surface higher. This is what, nowadays, overwhelms a coastal city unlucky enough to be in the path of a major storm—New Orleans when it faced Hurricane Katrina, for instance. But it is the tail end of the storm that leaves the final widespread signature across the sea floor. As the winds die, and the low-pressure zone fills, the mass of water settles back towards the sea. Made denser by the sediment swirling within it, sediment that was whipped up by the pounding waves, it flows seawards. This phenomenon—termed a storm ebb surge—flows out across the sea floor. As it slows and settles, it drapes a carpet of sand and mud, together with the remains of unfortunate sea floor creatures ripped from their homes (and these, of course, can bury and choke other unlucky creatures, that had survived the maelstrom of the heart of the storm). It is a new bed of sediment, that geologists term a storm layer or, rather more evocatively (and with a Shakespearean tinge) a tempestite. And it can stay as such, be buried by other such layers, and eventually become a rock stratum.

INTO DEEP WATER

Often, though, the sediment carries moving on, and on. It certainly did so in the case of the pebble stuff. One way to achieve this onward travel is simply not to stop. Or, the sediment can stop and settle for a few centuries or millennia, and can then be disturbed again, by a further storm, or perhaps by an earthquake. For we, and the pebble stuff, are now entering the realm which the wind and tides cannot reach, and in which gravity rules. And in this realm, finally, will be formed a part of the pebble that a child can see, without the aid of an electron microscope or a mass spectrometer, and even without a magnifying glass. *Now* will appear the largest stripes of that prettily striped pebble. The sediment grains are about to encounter what will be—for some 400 million years, that is—their final destination.

For gravity to do its work, it needs the wherewithal to act upon. This is provided by a slope, by the difference between high ground and low ground, or in this case between shallow sea and deep sea. Northwards from that ancient shoreline, the sea floor descended into depths of several hundreds of metres. Between the gently sloping coastal sea floor and the bottom of the deep sea was a slope sufficiently steep (not much is needed—perhaps only a few degrees) to render unstable any sediment accumulating on it.

Shake the ground around such a slope—with an earthquake, say, or an unusually powerful storm. Layers of the soft mud and sand can slide away, tumbling downwards. The steep scar at the back of this landslip is now inherently more unstable, and so further slices of sediment can slide away to reveal a new scar. This, in turn, then begins to collapse, which reveals a new back scar—and so on. In a modest event of this sort, many millions of cubic metres of sediment can be dislodged before the slope stabilizes (a cubic metre, to provide a domestic comparison, is about 25 standard builder's wheelbarrow loads). In larger events, billions of cubic metres can be removed, in what is a submarine catastrophe, albeit a commonplace one in the world's oceans.

Sometimes, the dislodged mass of sediment does not move very far. If the slope is short and relatively gentle, the sediment may simply pile up a short distance downslope, as a partly disaggregated mass of mud and sand containing dislocated slabs of strata; this kind of thing often happens with a landslide on land. But, underwater, much longer journeys can be undertaken, of up to hundreds or even thousands of kilometres. It is the subsequent travel of such material that is truly extraordinary.

It involves the making of a new kind of fluid: not quite mud, not quite water, but something in between. For as the soft stratal layers break apart, they become mixed with seawater to form a dense sediment-laden fluid, and it is this that, powered by gravity, can begin to flow downslope in what has now become a turbidity current.

Turbidity currents are amazing and hypnotic things to watch, even when reproduced at a minuscule scale in a water-filled tank in a scientific laboratory. The sediment-charged fluid moves across the tank floor, forever changing shape as it is distorted by the turbulent billows and vortices that are omnipresent in the current. It looks more like a living thing than a mere physical process. Once in this state, it can travel on extremely low slopes for long distances. Coming up against an obstacle, it can go around it—or even rebound off and travel in the other direction if the obstacle is big enough. The continually renewing turbulent gusts keep the sediment aloft above the floor of the tank—or of the sea floor—while the current is moving: thus, they are the key to the astonishing sediment-transporting capacity of these currents. (On Earth today, transport distances in excess of a thousand kilometres are not uncommon.) The mud, too, acts to 'thicken' the water—that is, to make it denser and more viscous, and this helps to keep sand grains also suspended within the moving current. All in all, these are highly efficient machines.

Even the most efficient machines eventually slow down. Once the turbidity current passes beyond the slope to move across a flat sea floor, the gravitational impetus that drives the flow, and powers those billions of turbulent gusts within it, eventually diminish and die. In the event recorded in our pebble, the current was slowing and becoming less turbulent, some 100 or so kilometres from its starting point (a short distance by global standards), and maybe a couple of hours after the triggering event high on the slope. Then, it shed its load of sediment across the wide sea floor—the heaviest particles first and then, successively the lighter ones. It took perhaps a few hours for most of the mud of this layer to settle onto the sea floor, though the finest material may still have been settling for days after the event, for at these depths neither waves nor tides made their presence felt. It was a still and—as we shall see—a strange sea floor to modern eyes.

Nevertheless, on this sea floor would settle a turbidite layer (and this process at least has not changed significantly since the Earth first possessed oceans). It is the grey stripe in the pebble: only a couple of centimetres thick, it is nevertheless part of something much bigger. Containing only a teaspoon or two of what-once-was-mud it is part of an event that took many millions of tons of sediment and spread them across hundreds—perhaps thousands—of square kilometres of the deep sea floor of ancient Wales.

It is a single layer, but it has neighbours. On that Welsh beach, with the pebble in your hand, you can turn to look at the cliffs and rocky crags behind you. They are conspicuously striped (Plate 1A). Each stripe is a stratum, an individual turbidite layer, mostly much thicker than the grey layer in the pebble, some reaching half a metre or more (I have seen one in Scotland recently, in rocks of about this age, that was some 40 m thick). They are what makes up most of the hills of this part of Wales, and much of the mountain ranges of the Earth.

One might walk up to the cliffs, and look closely at the stripes. They tell one story which cannot be seen in the pebble, or indeed deduced from any slab that is ripped from its original place in the cliff. Look for the thicker turbidite flow-units, those that carried a lot of sand (that is now sandstone) as well as mud, and then search for a stratal surface that represents the very bottom of the unit—in effect, the impression of a small area of sea floor (to do this, one often has to crouch or crawl—cautiously—beneath an overhanging sandstone slab). On those surfaces there will be—if one is lucky—curious markings. They are like short ridges with rounded tops with one end that gradually tapers into the surrounding slab and one that adjoins it steeply and abruptly. These had their counterparts—that is, elongated depressions—in the mudrock surface, just beneath the sandstone layer, the soft mudrock itself having been worn away by the sea and the weather.

These are called flute casts, and mark the scouring of the muddy sea floor by vortices within the onrushing currents that brought in the thick layer of sand and mud (Plate 1B). The tapering end points downcurrent, because the scouring vortex of sand-charged water always digs in steeply first, before dying out downcurrent. And so the steeper end points towards the source of the current. The flute casts on this coastline point consistently from the south—and so we know that the turbidity currents came from that direction. It is one of the signposts (a classic, now in every geology student textbook) that geologists use to reconstruct the geography of the Earth of the past.

THE CARBON SOURCE

There are other layers in the pebble with an entirely different history. The blandly grey thick stripes are interspersed with darker layers, mostly a few millimetres thick, that (especially when wetted) appear very finely striped. Look yet more closely (with a magnifying glass) and you will see that the stripes (geologists call them laminae) are less than a millimetre thick. Laminae of more or less the same colour and shade as the turbidite layer alternate with ones that are much darker—almost black, in fact.

A rule of thumb that holds pretty well for mudrocks in general—and for the pebble in particular—is that the darker grey they are, the more carbon they contain. The fine dark stripes are, therefore, more carbon-rich than the rest of the rock. Thereby hangs many a tale, and many a mystery. But we may start with how the carbon came to be in the pebble. We are dealing here with another journey, or rather a myriad journeys, for much of the carbon had little to do with Avalonia, and may have travelled from half a world away.

First things first. The thick grey layers that make up most of the pebble are turbidite layers that represent geologically instantaneous 'events'. Each time one was deposited, perhaps once every few decades, the deep sea floor received another thick carpet of mud and sand. Between these catastrophic events, sediment *almost* stopped arriving at the sea floor. Almost, but not quite. That sea floor, every now and again—perhaps seasonally—received thin dustings of mud and silt, the layers often being not much more than a grain thick. These came from slow drifts of muddy water that travelled across the sea, taking weeks or months for the journey rather than hours, the sediment slowly settling from them and falling into the depths. Slowly moving and dilute relatives of turbidity currents, these are termed nepheloid plumes. You can often see them if you look out to sea after heavy rainfall, as long trails of muddy water stretching out to sea from river mouths, and in modern oceans their slow progress and eventual dissipation can be tracked by satellite.

Interspersed with the nepheloid layers there is another slow downwards drift of material. But this is a track not of mineral grains, but of a continuous procession of death—of the remains of plants and animals that had lived as plankton high above, in the sunlit surface layers of the sea. Falling to the sea floor, they became the thin black layers that one can see (with the help of that magnifying glass) in the pebble.

The stuff of their bodies had not, for the most part, come from the neighbouring landmass, from Avalonia. The carbon had, say, been erupted as carbon dioxide from volcanoes from anywhere in the world; had then travelled around the atmosphere for some years; had then dissolved in the sea (as much atmospheric carbon dioxide eventually does), and in this form had been carried by ocean currents to the seas around Avalonia; and then been captured via photosynthesis by microscopic planktonic algae, which lived and died in these seas and—many of them—were eaten by planktonic animals, which lived and died in turn.

It is more of a world-encircling web, this, than the kind of geographically constrained pathways that were followed by the mineral grains. Nevertheless, at journey's end, both far-travelled carbon and parochial silicate come to lie side by side, at the bottom of the deep Silurian sea of Wales. It was an unfamiliar sea, for the world was different then. The pebble now can tell the story of the nature of this sea floor, and what that means for the Silurian Earth. For it was, then, an alien Earth.

The sea

THE DEAD ZONE

S ome things are just infuriatingly difficult to pin down in geology. For instance, just how deep was our pebble sea, the Silurian sea of the Welsh Basin at the spot that became, some 400 million years later, the beach beneath our feet? Well, one can estimate some kind of minimum depth. It was deeper than the depth to which waves and tides can leave a trace on a sea floor, because no traces of these phenomena have been found in the pebble stuff or—rather more convincingly as evidence—in any of the strata of those Welsh cliffs from which the pebble could have been derived. As a rule of thumb, that means that the sea was more than a couple of hundred metres deep, that being the depth to which the very biggest waves of the very biggest storms on a wide open sea can stir the sea floor.

Now, if strata have been deposited *above* that level, then one can make some reasonable estimates of ancient water depth. Thus, if one finds fossilized beach-strata, that is an obvious signal that those rocks were formed virtually at sea level. And below that, we can make a distinction between those shallow sea floors that are stirred pretty well all the time, even by the small waves of a fair-weather day (on this kind of sea floor, mud is winnowed away, and only sand and pebbles can settle); and those deeper sea floors only affected by the biggest storms (where thick muddy layers can settle in between major storms that may have been a decade—or a century—apart). But below even that? It is, in practical terms, hard to tell from the rock strata whether the ancient sea floor on which they were laid down was 300 m or 3000 m deep, or perhaps even more. So it is with the pebble rock. This Welsh sea floor was deep in general terms, but its precise depth remains a mystery—working out even a reasonably imprecise depth remains as a puzzle for future generations of geologists to solve.

But one thing can be said about it for sure, and that evidence can even be read from the tiny sample of it present in the pebble. It was a dead sea floor. Dead, that is, to any multicellular creature. Those delicate millimetre-thick laminae that slowly accumulated on the sea floor between the influxes of turbidite mud and sand are not just thin sediment layers: each lamina is also a canvas on which can be left the trace of any animal which walks across, or crawls through, or burrows into the sea floor.

No such traces interrupt the delicate layering of the pebble (Plate 2A). In the cliffs and crags from which the pebble was torn, there are thicknesses of these deep-water strata in which no tracks or trails are to be found. Not because such complex life had not evolved—because a wealth of fossil creatures—brachiopods, trilobites, molluscs—can be found in strata that formed in the shallower seas of these times. Something kept those animals out. This stratal signature, of an exclusion of life from the deep seas, is striking. It was noticed almost a century ago by one of the great pioneering geologists, John Marr. He said that the sea had been 'poisoned below the 100 fathom line'. And so it was, in a way.

Many factors can act to exclude living organisms. The water can be too salty, as in those landlocked seas where evaporation transforms seawater into concentrated brine—though here one normally finds layers of rock salt or, rather more subtly, the delicate impressions left of rock salt crystals that later dissolved away. Or the water can be hot and charged with chemicals near volcanic vents—though here one would expect to see volcanic rocks together with the sedimentary strata. Or the water may be literally poisoned, by toxic plankton blooms. But here, among the pebble stuff and its kin in the rock strata, there are no salt crystals, no volcanic rocks. And, the 'poisoned water' persisted at times for millions of years, not just in the ancient Welsh ocean, but also at the same time in seas in what is now the English Lake District and Scotland, and in Nova Scotia and North Africa and Poland too. There is only one way to kill off such an expanse of deep ocean simultaneously across so much of the world. That is to suffocate it.

These were oxygen-starved ocean depths. They allowed no form of life to colonize them, other than those microbes (of an ancient lineage, reaching back to the days of an Earth before plants and photosynthesis) that are adapted to anoxic conditions. This is quite unlike the familiar modern oceans, where oxygen—and complex multicellular life—persists virtually everywhere. Travelling back in time, in this case (as in many others, too) is like travelling to a different planet.

There are just a few representatives, on this modern well-oxygenated world, of the vast anoxic ocean depths of the Silurian. The Black Sea is the best-known one. It is deep—over two kilometres—and almost entirely landlocked. The surface waters teem with life (though pollution has lately been taking its toll). Below about 300 m, and extending right down to the bottom, the waters are devoid of oxygen, to the extent that they contain dissolved hydrogen sulphide, a gas well-known for its rotting-eggs smell, and slightly less notorious for its toxicity (it is, in fact, as poisonous as hydrogen cyanide). This is very bad for multicellular life but marvellous for marine archaeologists, because wrecks of Greek and Roman vessels, marvellously preserved, lie on the floor of this sea.

The open seas of today contain one or two tiny natural examples of such permanent anoxia. There is the Santa Barbara basin offshore from California, for instance. It is a depression in the sea floor, a few tens of kilometres across, surrounded by a rampart that prevents ocean currents from refreshing it. In those stagnant waters, there accumulate layers of sediment that show much resemblance to the layering in the pebble. Moreover, because those strata are still forming *now*, one can study them as they form, travel through the ocean depths in bathyscaphes to peer at them, take samples of those sea floor muds, leave open-topped jars on the sea floor to catch examples of the material that is falling onto the sea floor. It is a little playground, or paradise, for those trying to travel back to the past in their minds.

In the Santa Barbara basin there are thick layers of mud that resemble the thick grey stripes of the pebble. They are essentially turbidites, formed as major gravity-driven influxes of sediment triggered by major 'hundred-year' storms along the California coast. And the Santa Barbara muds also have, between the thick mud layers, intervals of darker, finely laminated mud. These modern Californian laminae, just like those of the ancient Welsh rock of the pebble, alternate between those that are very dark and full of organic matter, and those that are paler and resemble the material of the thick mud layers. At Santa Barbara, in the here and now, one can see that the pale laminae represent thin dustings of sediment that has drifted across from river floods each winter, while the dark layers are the remains of dead microplankton, fallen onto the sea floor. And these modern Californian laminae can be shown to be annual: one can count them backwards, down from the surface of the sea floor, just as one can with tree rings, and match up particular laminae with particular historical events—a particularly fierce flooding episode to produce a thicker lamina, for instance. It is a kind of calendar and modern history, combined.

No such luck with the pebble laminae though. The Santa Barbara basin gives a good picture—the best we currently have—of the general kind of setting in which the strata of the pebble were formed, and of the processes by which they formed. But it is not an exact picture, being strictly impressionist rather than a carbon copy. There is a devil in the detail of the pebble laminae that is yet another source of frustration to the budding student of this most ancient Welsh lore. For while the Santa Barbara laminae are essentially annual in nature and can be counted, charted, and analysed in these terms, the same does not seem to be true, alas, of the pebble laminae.

And here, one has to look at them *very* closely, either by cutting the pebble in half and polishing one surface, or by making a thin section of the rock for analysis by microscope. At this scale, the Welsh pebble laminae, so clear when held at arm's length, dissolve into a sea of uncertainty. Most of them, peered at closely, are not continuous laminae at all; rather, they are discontinuous wisps and lenses and patches, that often split into two, or grade fuzzily into other laminae rather than being clearly distinct from them. They are, in brief, impossible to count and analyse with any sensible level of precision. The frustration of it! I still remember the feeling of utter ... spifflication that I felt when I finally gave up the attempt at just such a lamina-counting exercise on a close relative of the pebble, some years back. For it would have been so marvellous, just so *cool* (I was young, then) to have, across a distance of 400 million years, an exact annual calendar of events in the Silurian. But it was not to be—and it still is not.

So why is the pebble not a tiny fragment of a precise stratal calendar, Santa Barbara-style? This is yet another mystery, born of the profound differences between the alien seas of the Silurian and the familiar oceans of today. In those chronically oxygen-starved Welsh seas, the chemistry was different, the current systems were different, the biology was different. The functioning of this world still remains, to us, largely obscure; much painstaking detective work still lies ahead, to enable proper illumination of the big picture of this segment of the Earth's deep history.

But there are also the minutiae, tiny details that need to be explained—and in this case the profound wispiness and fuzziness of those (in close-up) not-quite-laminae of the pebble. There is one clue in the open oceans of today. When dead plankton falls to the sea floor, it tends not to do so as the remains of individual organisms—for the bulk of the plankton is made up of single-celled creatures so small and light that it would take them a geological age (almost literally) to sink

THE SEA

individually through a few kilometres of ocean to land on the sea floor. Rather, they sink as clumps and aggregates, tangled in gelatinous masses, and—after having been eaten and excreted by larger organisms—as faecal pellets. Marine snow, these falling clumps have been called, and the 'eternal snowfall of the oceans—as Rachel Carson evocatively put it—comprises a kind of expressway from the surface waters to the sea floor. In highly productive areas of today's oceans, one can look out from the windows of a bathyscaphe and see exactly such a snowfall drifting downwards. Thus, some of the wisps one can see in the pebble might be, quite simply, the remains of individual flakes of Silurian marine snow.

That is possible. Even plausible. But, as ever in science, there are other possibilities. There is the exact nature of that alien sea floor to consider. For, more likely than not, that sea floor would not have been just a flat expanse of mud and silt and fallen marine snowflakes. There would have been life there too, adapted to those eternally dark and oxygen-free spaces: the ancient microbes of the anoxic realm. Now microbes are commonly regarded as the simplest and most primitive of life forms that live in soils and compost heaps and such; and when they chance to live in *us* too successfully, we call them germs.

That hides an increasingly appreciated sophistication, and a sheer skill at *togetherness*. Many microbes are colonial, and their colonies are amazing (one can almost see the raised eyebrows of the microbiologists as they write about them). The colonies are everywhere, from the skin of one's teeth to the scum of a pond, for example. Each can involve dozens or hundreds of bacterial species, with the individual bacteria signalling to each other chemically (quorum sensing, it's called) as to which of their fellow microbes to let in and which to keep out, coexisting

and co-operating and co-evolving in a delicate set of associations and alliances—and indulging also in inter-microbial
colony warfare when necessary. They are microscopic equivalents of a rain forest, or a coral reef. In modern times, the sea
floor is a tough place for them. Left to themselves, the microbes
would cover pretty well every submarine surface with mat-like
colonies. Unfortunately for them, they are also edible. Their
giant enemies and predators, the worms and grazing molluscs,
are continually ripping the complex and delicate microbial
mats to shreds as they burrow and chew their way through
the surface sediment.

The widespread anoxic sea floors of the Silurian were no
place for grazing worms though, and hence they could revert to
being the undisturbed haven of the microbes (as they had been,
indeed, since the dawn of life over 3 billion years ago until the
invasion of the grazing metazoans 2.5 billion years later). This
would have been a microscopic world of staggering richness
and diversity, of battles and treaties and alliances, and of
treachery too. All this would have been mediated through
quorum sensing (and probably by other means that we huge,
lumbering bipeds have not yet had the wit to uncover), much as
we conduct our contemporary business by email and mobile
phone, as different groups of microbes struggled for ascen-
dency on a terrain and battlefield that stretched all around the
Earth.

And all that is left of this richness is the suspicion that
some—perhaps many—of those carbonized wisps in the
organic-rich laminae of the pebble were, at one time, the
remains of microbial colonies, in whole or in part. It is yet
another example of how incomplete, still, is our understanding
of the past—and particularly of the things that *matter*, because

the microbes then, as now, ruled the world—and how much there is still left to study.

DISTANT ICE

The laminated organic-rich mudrocks are one major and highly distinctive component of the Silurian strata, and are fairly represented by our pebble. But they have a direct opposite, a stratal *alter ego* that may be seen in many sections of strata in the cliffs—or might have been encountered if one had picked up another pebble. This alternative rock type does not possess the fine dark striping between the smooth thick layers of turbiditic mudrock. Instead, in this position, it has exceedingly pale-coloured layers that generally look a little nondescript, except that, scattered here and there, they include small dark streaks and spots, easiest to see when the rock is wetted (Plate 2B).

These streaks and spots are, when closely looked at (and when one converts the two dimensions of the rock surface into a three-dimensional pattern) are fossilized burrows of some sort of organism. Of what, we don't know, but a default interpretation is that they were worms of some sort. Whatever produced them, though, were mobile multicellular creatures, and this clearly shows that the Silurian sea floors were not always anoxic, but rather alternated between a state of anoxia and one where there was sufficient oxygenation to allow colonization by burrowing animals—with the disruption they caused to the fine sedimentary layers.

The amount of burrowing, in fact, must have been intense, for typically no vestige of the delicate laminae survive, while the very pale colour of these layers is because most of the organic matter in them has been 'burnt' off, used as an energy source by all the oxygen-adapted organisms (including different—now aerobic—assemblages of microbes) of those sea floors.

Hence our pebble, with its fine dark laminae, represents just one half of a pair of contrasting ocean states of the Silurian, which alternated with each other as geological ages passed. Each state could persist for millions of years (the first 5 million years or so of the Silurian in Wales was almost continually anoxic, for instance). At other times the oxygenic and anoxic states could flip between each other every few thousand, or tens of thousand, years. It is a pattern that is so deeply imprinted into, and so fundamental to, these rock strata that it is the basic means by which the strata are subdivided on the geological maps of Wales.

But what was the cause, and significance, of this phenomenon? Why should an ocean switch from being oxygenated and widely colonized by multicellular animals to become a killing zone for these creatures, and a paradise only for anoxic microbes? Here again, we enter the realm of mystery, in contemplating the mechanism of a distant, vanished world. It is sobering to recall that we only partly understand the functioning of our modern world, where sea and land and atmosphere can be continually monitored, sampled, and analysed—witness the way in which, say, some symptoms of global warming (polar ice melt and such) are currently advancing at a much faster rate than predicted by the latest computer models. Nevertheless, there are some tantalizing patterns in these fragmentary messages from the Silurian world, and some basic deductions that can—cautiously—be essayed.

For instance, there are really only two ways in which an ocean can be turned anoxic. One way is by producing so much organic matter that all of the oxygen supply is consistently exhausted in trying to oxidize it. This is basically what happens when, say, too much agricultural fertilizer pours off farmers' fields into rivers and lakes: this stimulates the growth of so much algae that, as these die and decay, the oxygen is used up, and the rivers and lakes go stagnant, killing the fish in the process. It is a common problem today. The second means is by holding the nutrient levels constant, but instead cutting off the oxygen supply. This is what effectively happens in the Black Sea—the upper oxygenated layers are less salty and so form a low-density 'lid' on this sea, preventing the circulation of the waters that would allow oxygen into the deep waters. These two methods are not, of course, exclusive: one can have some measure of each of them, combining to starve a sea of oxygen. Which of these might have been of most importance in the Silurian seas of Wales? It is early days, again, as regards this question too. (Remember that geology is a *young* science—it encompasses so much history, and there is still so much to find out.) But there is one intriguing connection.

The major episodes of anoxia in the sea coincided, more or less, with times when sea level was high, while times of oxygenation took place when sea level was low. This can be established by tracing the strata across country and matching up the timing of a change in deep-water oxygenation state in one place with, say, a shift in the position of the palaeo-coastline elsewhere. It's a painstaking science, this matching up of the timing of different events in different places, and it depends on slow and careful fieldwork and the exact dating of strata by their fossil content. Nevertheless, enough evidence has accumulated to equate rising sea levels with oxygen starvation at depth. How much did sea levels rise and fall? An even trickier question, this, with rocks that are so old, but a few tens of metres would be a reasonable estimate.

The best way to raise and lower sea level quickly and substantially is to alter the amount of ice in the polar regions. As the Earth chills, and icecaps grow and the glaciers advance, water is drawn out of the oceans. Then, when the Earth warms, the ice melts, pouring the water back into the oceans. This is what has taken place in the Quaternary ice ages that we still live in (though we collectively are altering the Earth's climate controls on such a heroic scale that this situation might not last for much longer). A similar process seems to have taken place in the early part of the Silurian Period and in the latter part of the Ordovician Period that preceded it. The end of the Ordovician, indeed, has long been known to have been marked by a brief, but intense glaciation that devastated marine life, with ice growing and then decaying on South America and southern Africa, which were then joined together and positioned at the South Pole.

But this may not have been all. Evidence has been growing that the ice waxed and waned on a less dramatic scale for some ten million years before and after that climactic event. This seems likely (*plausible*, perhaps would be a better way of putting it at our current state of knowledge) to have been what drove sea level up and down across the world. And the changes in sea level, then, look to have also led to the suffocation of large parts of the deep seas of the Earth (and, as the most minor of by-products, the fine lamina-striping of our pebble)—and their subsequent re-oxygenation.

Does this help us to choose between our two models for oxygen starvation—over-fertilizing the oceans or cutting off the oxygen supply? Perhaps. One of the reasons why we

81

currently have a healthily oxygenated sea floor (not so health-ily, though, for those microbial mats) is that there is a vigorous ocean current system, driven by the world's polar regions, which is distributing that oxygen. In those regions, as sea-ice forms each winter, the surface water that remains is cold and salty (because the sea-ice excludes salt); it is therefore more dense than the water beneath, and so sinks to the floor as the motor and the initiation of the world's ocean current systems.

In a warmer world, with higher sea levels and smaller ice-caps, this polar motor for ocean currents should be weakened, and hence the currents should be slower, thereby reducing the rate at which oxygen would be supplied to deep water. Con-versely, when the icecaps grow, the polar motor should strengthen, thus bringing oxygen flooding back to the deep waters. This is at least moderately plausible—for the time being. It does not exclude models of oxygenation based on nutrient supply, not least because too little is known about what controlled marine nutrient levels in the Silurian. But it is some sort of start in the long (never-ending, I fear) process of reconstructing how this long-vanished world truly functioned.

At the risk of stepping out yet further towards the end of a plank that stretches out above a bottomless ocean, there is a further facet that we might consider: that the pebble does not merely bear the mark of a particular ocean and climate state, but that it is one minuscule part of a planetary machinery that stabilized the Earth's climate, and that kept the Earth from overheating. The basis for this idea is simple. The dark laminae in the pebble reflect concentrations of carbon, which was being buried in the muddy strata of the stagnant, anoxic seas and thus was prevented from being released back to the oceans, and subsequently to the atmosphere, as carbon dioxide. As more and more carbon was buried, the level of carbon dioxide in the atmosphere would begin to decrease. This, again, seems plausible, as the carbon-rich anoxic strata can form units many metres thick that extend over many thousands of square kilometres; the amounts of carbon involved eventually run into many billions of tons. (If we as a civilization could bury only a fraction of that amount in our proposed—though as yet hypothetical—carbon sequestration programmes, then global warming might indeed be effectively mitigated.)

As the carbon dioxide levels began to fall, so global temperatures would begin to drop. Ice, then, would begin to build up in the polar regions; sea level would fall, and the ocean circulation become more vigorous. This would switch the oceans to an oxygenated state, and the carbon—those countless dead plankton—that was hitherto being buried would instead be consumed, respired, oxidized, and return to the oceans and thence to the atmosphere as carbon dioxide. As this continued, atmospheric carbon dioxide levels would begin to rise again, until the Earth warmed sufficiently for those ice caps to melt, and sea level to rise again...

It is a nice hypothesis, a neatly symmetrical way of keeping the Earth's climate in balance. A former planetary thermostat, if you like, of which the pebble is a tiny part. Or, more technically, a negative feedback mechanism, a means of preventing both a runaway greenhouse and a runaway icehouse. Plausible, yes—particularly as this was an Earth in which land-based vegetation and terrestrial humus-rich soils had yet to develop, and hence an Earth on which the oceans likely called the shots as regards climate. But true? Or, more realistically, true enough to have had a significant influence, among a complex array of competing factors, many of which we are blissfully unaware of (or perhaps, because we are scientists: *fretfully* unaware of)? Like many other nice hypotheses in science, it needs testing, to destruction if need be, by putting on one's field boots and collecting more data. Eventually, that nice hypothesis will be supported, strengthened even; or modified (perhaps beyond recognition); or simply discarded. That's the nice thing about science: it'll all come out in the wash.

BACK TO THE PAST

There's a final coda to this journey through the pebble, though it's not a particularly heartening one. The world of our pebble, here, is of a sea floor devoid of the life we most identify with, the life of multicellular organisms, of marine creepy-crawlies of diverse sorts. It is a profoundly different world from our own. Yet, there are signs that, in places, it might be returning, because of human activity. This is not (yet) because global warming has eliminated so much sea ice that the Earth's ocean current system has slowed. The reason is more prosaic. The nitrates and phosphates applied to the land to grow the crops we need to live on are being washed off that land and down the rivers and into the sea. In those seas, the fertilizers are stimulating massive growths of plankton and algae that are beginning to cause an oxygen deficit in those seas, causing multicellular organisms on the sea floor to suffocate *en masse*.

Dead zones, they are called. There is one in Chesapeake Bay in the United States, where the River Susquehanna meets the sea. There is another in the Gulf of Mexico around the mouth of the Mississippi; and yet another in the Baltic Sea. They now cover thousands of square kilometres, though to us landlubbers they are still effectively invisible, and generally out of mind (imagine the uproar if everything bigger than a bacterium was killed off over an entire county, for instance). They are still mostly seasonal: that is, the killing takes place in the summer months, when the oxygen deficit is highest, and an increasingly frazzled ocean floor biota tries to re-colonize in winter.

We are not yet in the pebble territory of perennially and permanently oxygen-starved seas, Silurian style. Nevertheless, the pebble is a reminder that the Earth's oceans are capable of existing in states very different from the one we are familiar with. The pebble also holds, within its small volume, evidence of the life of those long-vanished oceans. Unlike the ocean states, the very particular organisms of the Silurian will not reappear in some post-human future. Now long gone, never to return, these bizarre organisms nevertheless testify to the diversity of forms that life can adopt. It is time to explore the ancient biology trapped within the pebble.

Ghosts observed

UNDER THE MICROSCOPE

Life is ubiquitous on the Earth's surface. Exuberant, fantastical, tough, and very, very persistent, it gets pretty well everywhere. Darwin marvelled at what could be found in a simple tangled thicket by a footpath, while a spadeful of soil can keep a zoologist occupied for weeks—all those mites and worms and springtails and leatherjackets—and a microbiologist busy for months. There is life in the hottest deserts and in Antarctic ice and nestling up against boiling volcanic vents. It flies high through the air too—not just birds and bees, but spores and pollen and aerial bacteria (so abundant that they can make rain fall more copiously by acting as nuclei for the raindrops).

In death, too, the organisms can be tough. Not every corpse gets recycled back to form new generations of the living, and not all fossils are such scarcities that each becomes a museum piece or commands a handsome reserve price at an auction of ancient curiosities. The ghosts of the past are all around us, in solid form. Indeed, we owe to them the comfortable contemporary life (not enjoyed by all, admittedly), of centrally heated houses and easy travel and an abundance of food. The remains of dead plants and animals power contemporary human civilization, in the form of oil and gas and coal. At a price, of course, that is still to be paid.

The pebble contains a little coaly stuff within it—tiny flecks of what is now essentially carbon, which gives the dark laminae their colour. It probably makes up, today, something over one per cent of the pebble; when the pebble stuff had been layers of mud and sand on that Silurian sea floor, it would have been nearer 10 per cent. That carbon was once living things—but how does one go about finding what kind of living things these once were?

The easiest way to release the cornucopia of ancient life locked in the pebble might strike a disinterested bystander as a little harsh. Indeed, it would be quite terminal for the pebble, albeit highly revealing. The procedure is, by now, quite standard. The pebble (or its neighbour) is crushed into fragments and put into hydrofluoric acid that will dissolve rock but not tough acid-resistant fossil material. What is left is sieved, washed and put on a microscope slide.

Let us look at the smallest material first. Under the microscope will appear a mass of black fragments of various shapes. Many will look just like microscopic coal fragments, with no particular resemblance to any living organism, or to cells or bits of tissue. These are what is often called 'amorphous organic matter' by the scientists—termed palynologists—who study such fossil material. It was once soft living tissue, perhaps of the microbial mats, or flakes of marine snow, but it has subsequently been so mulched and squashed and heated that no trace of the original biological structure has remained, except a residue of carbon. There's much of interest in the squashing and heating, as we'll see later, but it has obliterated most of the detail of the life of the past.

Not all, though. Amongst these obscure and shapeless fragments there are more distinct shapes. Tiny black spheres, for instance, a few tenths or hundredths of a millimetre across. Some are simply more or less smooth spheres, a little like microscopic blackened billiard balls. Some have thorn-like outgrowths, either as single sharp points, or with their ends split into strands and filaments. Some are approximately pyramidal in shape. There are thousands of these objects in the pebble.

They do have a name, although naming something doesn't always convey much in the way of understanding. The name itself is apt—they are called acritarchs, which means 'things of confused origin', the term being applied to any more or less tiny round-ish hollow fossil objects that have an organic wall tough enough to resist hydrofluoric acid. They probably represent organisms of different types, among which may be large single-celled green planktonic algae. But these acritarchs are not the algae themselves—they are the cysts of the algae, hard outer protective coatings that develop when times are hard for one reason or another—when food is scarce, say. It is a kind of hibernation mechanism. When the algae grow such tough coats, they sink to the sea floor and wait until conditions get better, emerging through a split in the casing if it looks as though the good times are beginning to roll again, and heading back up to the sunlight.

FIGURE 3 Silurian microfossils, of the kind that one might find in the pebble—an acritarch (top) and two chitinozoans (bottom).

Today, the likely descendents of the acritarchs (of some of them at least) are mobile marine algae called dinoflagellates. These also grow cysts, which have a rather more sophisticated design, including a nicely engineered escape hatch. Some, like the formidable *Pfiesteria*, have the sinister reputation of being able to multiply enormously to produce 'red tides', toxic blooms deadly to fish (and to people too). Dinoflagellates seem to produce cysts especially plentifully in the aftermath of such blooms, and so the acritarchs in the pebble may well be an echo of similar booms and busts in the Silurian. Perhaps the booms were also toxic. Chemical warfare is a simple and effective means of settling inter-species disputes; it probably has a very long history on Earth.

Among the acritarchs in the pebble, I'd hazard a guess that there would be a few specimens of one type called *Moyeria*, which is shaped a bit like a tiny deflated rugby ball. *Moyeria* is thought to have lived in fresh water, but it was easily washed from the land by rivers, and carried farther out to sea by those efficient turbidity currents. It is probably a Silurian relative of the protozoan *Euglena* that abounds in ponds and streams today. *Euglena* is a stalwart of school biology lessons, almost as much as is the amoeba. It is not quite plant (it hunts, kills and eats other micro-organisms) and not quite animal (it has chlorophyll, and photosynthesizes). And it also forms cysts in bad times. One can imagine it starring—once it's enlarged somewhat—in one of the scarier Hollywood monster movies. Some forms of *Euglena* move towards the light, a form of behaviour studied at—and I quote the address verbatim—the Laboratory of Marine Serendipity and Spectaculars in British Columbia (it's a workplace address that one confesses to a deep envy of). Those *Euglena* that show such behaviour are marked with a particular spiral pattern—and such a spiral pattern has been reported from fossil *Moyeria* too, suggesting that it too swam towards the sun.

There are larger fossils, that appear among the remnants of an acid-dissolved pebble. That material, separated on a coarser-mesh sieve, will show dozens—or perhaps hundreds—of little flask-shaped objects, each about the size of a pinhead. These are also made of tough organic material, which was thought to be chitin, the stuff of which the exoskeletons of crustaceans and insects are made, by the man who discovered them in the 1930s, the German (East Prussian, then) Alfred Eisenack. Eisenack was a man whose passion for research was interrupted, but not blunted, by two long spells as a prisoner of war. After the First World War, he described his captive sojourn in eastern Siberia, where he worked as a chemist, as the best spell of his life. However, of his second, long bout of captivity, after Red Army soldiers had upended him, dragged his boots off, and forced him to march to the Soviet Union, he subsequently said little.

The chitinozoans are probably not made of chitin, as no trace of this substance has been found by more sophisticated chemical analysis. It's still not known quite what the material is, because the passing millions of years have not been kind to the complex and delicate organic molecules (although they did not weaken their resistance to the fierce laboratory acids). The name 'pseudochitin', that is 'looks-like-chitin-but-probably-isn't', has been applied for want of anything better. But, as ever with fossils, the shape is the thing.

Chitinozoans are essentially little bottles: some large, some small, some with flat bottoms, others being rounded. Some have, like the acritarchs, spine- or thorn-like projections in various arrangements. The surface of these flasks, looked at with a scanning electron microscope, has a wonderful variety of textures, which has tested the descriptive powers of palaeontologists to the full. For collectors of rare and precious examples of the obscure British adjective, there is a goldmine here. Scabrate, for instance—that means with irregular wrinkle-like ridges.

Or vermiculate—that is worm-like, when the ridges become more sinuous. Or foveolate—here the surface has a network of raised ridges in a honeycomb pattern. Or verrucate—well, parents of young children will have no problem with *that* one.

It means that there is a lot here to enable the description, and categorization, and analysis of these objects. There is a lot to go on, plenty of data. But, all description made: what are, or were, the chitinozoans? How can one interpret such miniaturized fossilized bottles?

This is still a puzzle—far more so than with the acritarchs. There have been many suggestions. Chitinozoans have been thought to be whole organisms in entirety—even overblown relatives of the acritarchs, that is, plants. Some have thought them to be fungi, and others amoebae. The hypothesis that is way out in front, though, and has been since their discovery, is that they were egg-cases of some larger organism. There is a slight drawback to this, in that there are no egg-cases remotely like the chitinozoans today. Nevertheless, there is some evidence for this view, albeit perhaps a little on the circumstantial side. They are not always found as individuals. In places, they occur as chains—or, more remarkably, as large, spirally arranged clusters, arrangements associated with the laying of multiple eggs.

But eggs of what creatures? Some of the organisms that are also common in Silurian rocks have been suggested. The orthocone nautiloids, the straight-shelled ancestors of the ammonites, for instance; or the cystoids, relatives of the starfish. Alas, these ideas don't quite work, not least because while these are sometimes found in rocks of the same type and/or age as have yielded chitinozoans, they are not found together consistently *enough*. There's always a mismatch sufficiently big to prevent any of the other well-known candidates (trilobites, say) being regarded as plausible chitinozoan-producers, while others (the annelid worms) are too rarely fossilized to make direct comparison with.

So, there was a mystery organism out there. It has been reconstructed from what is known of what chitinozoans are and where they are found. Thus, the animal (it doesn't work as a plant) was multicellular, somewhere between a few millimetres and a few centimetres long. It was entirely soft-bodied—other than its eggs, which developed inside it and were released, fully formed. It floated, or swam in the sunlit surface layers of the sea, because the chitinozoans are spread widely across strata that represent both shallow and deep Silurian seas, including those that were starved of oxygen at depth. It had certain dislikes—it avoided coral reefs, for instance. This phantom animal is, thus far, just a set of inferred attributes and habits and qualities; it doesn't yet have a form, or a face, or a position within the tree of life.

Out there, somewhere, will be the Rosetta Stone of the chitinozoans, a slab of rock that will preserve both the carbonized impression of the chitinozoan animal and the chains of eggs within it, waiting to be released. Someone will find it, and realize what it is (that second part is crucial, because it is possible—perhaps even likely—that the crucial slab lies in a museum vault somewhere, never yet stumbled upon by a palaeontologist with the necessary expertise). Then, there will be a brief fanfare among palaeontologists, and celebrations and the concoction of a purple-prosed press release. Another enigma of the fossil realm will have been resolved, and another part of the Silurian sea will come into focus. And the next day, palaeontologists (one or two with slight hangovers) will get back to work, for there is much more work to do out there, and many more mysteries to be solved.

THE HOUSE-BUILDER

But that is not the end of the pebble menagerie. There is another creature in there, with a much longer pedigree (of study, at least), a form that is wonderfully distinct *and* distinctive, and with, as we shall see, a modern relative so close that it might be a living fossil. Yet, paradoxically, this other plankton creature vies with the chitinozoans as regards mystery. Indeed, it might as well come from outer space, for all the resemblance that it has to any modern member of the plankton.

Look closely at the pebble, among the fine dark stripes, and, here and there, there is a gold fleck, perhaps admixed with some reddish-brown, about a millimetre across. Look closer, with a hand lens, and this fleck is surrounded by the narrowest of black rims, and beyond that there is a wider rim of a pale fibrous mineral. At this point it is time to get to work. To uncover this monster of the deep (and to the microplankton, it would have been truly monstrous), one cannot use scarily powerful acids—not from our pebble at least. The monster would not dissolve, but it would crumble into tiny fragments on being released from the pebble. The approach needed here is akin to brain surgery, but on rocks.

You take a sharp needle mounted in a pin-vice, which you can hold like a pencil. An ordinary sewing needle is somewhat too springy—what is best is an old-fashioned steel gramophone needle. The work needs to be done under a binocular microscope (a hand lens is quite insufficient here): one needs both hands, strong magnification and a good light. And patience. Holding the pebble in one hand, you press with the pin at the rock just above the black-and-gold fleck. You need to press hard enough to break a sliver of rock off the pebble, away from the fleck—but not so hard as to go through into—or indeed to touch—the fleck itself. In practice, that means that both sets of muscles in your arm need to be flexed in opposition, one set to push the needle down, and the other to hold the needle-point back at the *exact* moment that the rock sliver breaks away. It needs a good deal of care to do this precisely, so that the black material—a small fraction of a millimetre thick and very friable—is not dislodged as well. If it is, then the fossil material itself is gone, shattered into tiny fragments. But all is not lost; the golden material just underneath is tougher, and it provides an effective facsimile of the interior of the fossil.

You proceed, removing the rock, sliver by sliver, and more often grain by grain, from the black-and-gold fleck. After a few minutes, the fleck, now being granted the third dimension, reveals itself as a tubular structure going into the rock, a very delicate black pipe with a solid golden infill. The pale fibrous mineral just above it is sacrificed as well, to reveal the black material underneath it (though those mineral fibres will eventually have their own uses, as we shall see later). A few minutes later, you realize (your arm will be tiring by now, from the effort of keeping the muscles tightly flexed and simultaneously under minute control) that the tube is widening—and then it very abruptly narrows. As you dig farther back into the rock, it slowly widens again—and then suddenly narrows once more. Enough! It is time for a rest, and a cup of tea.

What you have revealed is a structure that looks a little like a slender fretsaw blade. If you carry on pursuing it back into the pebble (excavating a deeper and deeper channel into the rock as you go), it may carry on for a centimetre, or more. It may widen, or narrow, or flex gently, or curve sharply, or loop around in a spiral, or the pattern of the saw teeth may change from low- to high-angle. The geometrical possibilities are many. It is a graptolite (Plate 2C-F, and Figure 4).

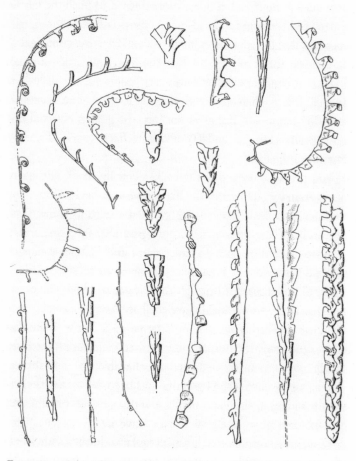

FIGURE 4 Colonial patterns in the Silurian plankton—camera lucida drawings of fossil graptolites, magnified about six times.

Graptolites are strange-looking fossils, looking at first glance more like some complex hieroglyphics or crystal growths than the remains of what were once living organisms. They have had a poor press. As regards fossil star quality and popular appeal, alas, they trail not only well behind the dinosaurs—as does every organism living or dead—but also behind the trilobites and ammonites and even the humble sea-shells. Among the more juvenile students of geology (of whatever age), graptolites have a reputation for: first dullness and second (as an antidote to the dullness) of being reasonably reproduced on a rock surface with a pencil in a moment of artistic inventiveness, to the delight of the prankster and the confusion of the poor benighted teacher.

And yet, they are a biological enigma of the first order. Most fossils can be interpreted by comparison with living relatives, to make their reconstruction, if impressive—in the case, say, of a *Tyrannosaurus rex*—at least familiar. *T. rex*, with its sharp teeth and legs built for bursts of speed, was probably a top predator—or, in some interpretations, an efficient and mobile scavenger. Using the behaviour of modern ecological equivalents (crocodiles, say), one can reasonably model them and their behaviour for that Hollywood blockbuster or that TV docu-drama. With the graptolites, there is a living relative—and yet it is one that makes the interpretation of these extinct creatures astonishing, and to some scientists well-nigh unbelievable.

First things first. The object that has been so carefully excavated is essentially a tube, one filled with a golden mineral (and *that*, of course, will also eventually have its story). The black tube is not a simple one, but has regularly spaced offshoots—about every millimetre—that open to the outside world. These are what appeared as the individual saw teeth of that fretsaw-blade-shaped fossil. The tube is made, like the preserved cases of the acritarchs and chitinozoans, of a tough organic substance. Now blackened and carbonized, it is thought originally to have been collagen, the material of which, for example, our fingernails are made. The whole thing was a colony, with each of the short side tubes being a 'house' for an individual animal, or zooid, with these individuals likely being joined along their common corridor, like a team of climbers roped together. These colonies were part of the plankton, and like the acritarchs and chitinozoans, fell onto the anoxic sea floor after death and were entombed in those stagnant black muds. Looked at closely, their overall shape is remarkable, a structure that could achieve baroque complexity, and nevertheless remained, for each species, precisely and consistently engineered. Well-preserved graptolites (even the ones in the pebble, if you're lucky and have been careful in your work with the needle), show that the tubes, however complex in overall shape, are made of many successive rings, and these betray the pattern of growth, from the early beginnings to the mature colony.

Present on the sea floor today, here and there, are clusters of tubes of vaguely similar dimensions, also colonial, also made of a tough organic substance, with each tube also made of successive rings. You can find them, say, on the bottoms of some Norwegian fjords, or off Plymouth Sound, or—and this is where you might be more tempted to check them out—around the islands of the Bahamas. In each tube is a tiny animal that, when it shows itself to the world, is rather beautiful. It is a blob of tissue, surmounted by an array of delicate tentacles; it feeds using this impressive device, filtering microscopic floating organisms out of the sea water around it. This organism is a pterobranch (meaning 'winged arm') and if you have not heard of it previously, you are in large company, as it is one of the more obscure bit players of the biology of the modern sea floor. Because of the similarity of the ring-structure of its tube to the fossil graptolites, it is a very good candidate to be a living relative of the graptolites, despite the fact that it skulks on the sea floor (often hiding under upturned sea shells and the like) while the graptolites once lived—and maybe ruled—the high seas.

But the pterobranchs, although shy, are remarkable in one respect. They are architects and builders. Their tube is not a skeleton, as is the shell of a mollusc, or the bones in our bodies, or the hard calyces of corals. It is a construction, much as is a spider's web or a termite's nest: a product of animal behaviour. Beneath a pterobranch zooid's mass of tentacles, near its mouth, is a round disc of tissue. Every so often, the creature stretches this disc out, clamping it over the end of its tube. Unclamping it a few hours later, the tube has been made longer by one newly added ring; and so the home has been extended, and the zooid can make itself that bit more comfortable.

Did graptolites do the same? They were once thought not to, as the constructional problems seemed too formidable for such 'lowly' organisms to overcome. For while the pterobranchs do a perfectly effective job, it is (one almost needs to apologize on their behalf) little more than jerry-building, for their tubes are an untidy mess, more akin to a small tangle of spaghetti than anything else. The graptolites, by contrast, are marvellously engineered, a Fabergé egg against the child's clay model of the pterobranchs. Not only is each individual's domicile wonderfully structured (often with lids and flanges and thorn-like projections), but each of these chambers is joined seamlessly to the adjacent one, showing perfect co-operation between the neighbours in achieving the join.

Impossible? Well, when electron microscopes were first used to analyse extremely well-preserved graptolites, attention was focused on a surface layer that covers the rings just as plaster covers the bricks on the wall of a house. This surface layer was seen to be made of variously arranged flat strips of hard organic tissue, reminiscent of bandages very untidily wrapped around an Egyptian mummy. There is no sensible way that such a layer can form as an internal skeleton beneath soft tissue, and the only reasonable interpretation is that the zooids, once they had finished manufacturing the basic shell of their home, then finished the job by literally plastering their construction inside and out, sweeping their secretory discs across the wall exactly as a plasterer wields a trowel. Nevertheless some scientists still shake their heads, and point to some outrageous design—the kind of graptolites, say, that don't have solid walls but a mesh-work, that the impossibly skilful zooids must then have well-nigh *knitted*—and say that it couldn't have been so.

For what it's worth, I think that the evidence in favour of graptolite zooids as builders is so compelling that it probably *was* so. But it is still so fantastical an image as to seem almost science fiction: these bizarre twig-like colonies, building their own homes that doubled as ocean-going vessels, and then sculling across the sea in them. Mind, another of the ongoing debates is as to whether these colonies floated, or whether—by synchronizing the activity of their beating tentacles—they swam.[4] (And the tentacles that we give them in our imaginations are directly borrowed from the pterobranchs, for these delicate

---

[4] And there has also been a debate whether such plankton can, by the beating of their tentacles or flagellae *en masse*, effectively *mix the ocean water* itself. It's a lovely idea, and according to recent research, perhaps even with some truth to it.

structures have never been found fossilized.) Nonetheless, one can imagine the graptolite creatures feeding on the acritarch algae; and competing with, or ignoring, or chasing, or hiding from the mystery chitinozoan creatures; and then breaking off to perfect their architectural skills. A wonderful life, indeed.

It's marvellous getting a glimpse of this Silurian sea, so unlike our own. No fish, no dolphins or whales, no porpoises or seals or walrus. A world, instead, of small plankton, whose petrified almost invisible remnants are now packed into the pebble, the lives of which have been painstakingly reconstructed by many years of work by many scientists. Yet, these hard-won glimpses have not, for the most part, been driven by simple curiosity, although that certainly had a part to play. There is a strictly utilitarian reason for wanting to find out as much as possible about these once-living organisms. For, in containing these thousands of corpses, the pebble not only contains a record of biology and ecology. It contains time itself.

THE CAPTURE OF TIME

Time, here, is not some abstraction for philosophers and cosmologists to ponder on. It is the tool by which geologists convert what is otherwise a planet-sized mass of incomprehensible rock soup into a coherent geometrical pattern and then from *that* read a sensible history, a narrative of Earth events. It is nonsensical to take, say, 100-million-year-old preserved lakes in China, 200-million-year-old river strata in Russia and 300-million-year fossilized delta deposits in Germany and, from that, to reconstruct an ancient landscape. To produce a snapshot of the ancient Earth, one needs to make sure that the different parts of the picture represent the same time interval, as nearly as possible.

And the best practical guides to deep time are still fossils, for each biological species or group that has ever existed on Earth has had a time of origin, a time of existence and a time of extinction. And it didn't come back.

The slicing of time into units is easy when the time units are very big. Thus, the planktonic graptolites in Britain are only found in rocks of Ordovician or Silurian age, that is, in strata that we now know to be between about 500 million and 400 million years old. Find even one fragment (that even a first-year geology student should be able to recognize) and the strata from which it came must be of that age. So far, so good, but to subdivide that large unit of time, one needs to work out that there were different species of graptolite that lived at different times—and then to learn to recognize those different species and to know just when in geological time each one lived—and when each one died out.

It was Sir Walter Scott who, entirely unconsciously and indeed posthumously, caused the graptolites to be recognized as superstars of geological time, well over a century ago. This inexhaustibly prolix author was a favourite of Charles Lapworth, a bookish child in Faringdon, Bucks. Lapworth became a schoolteacher, and chose to work in Galashiels in southern Scotland, because that was the countryside of Scott's romances. The spirit of Ivanhoe did its stuff, and Lapworth found and married a local girl. He also found another love, for what was to become a lifelong relationship. Having taken to walking in the hills of the Southern Uplands of Scotland, he noticed some thin layers of mudrock that bore the hieroglyph-like markings of graptolites. That was a bit of a find in itself, for the Southern Uplands are mostly—99 per cent or more—made of coarse sandstones that are desperately poor in fossils.

He began to take an interest. I would love to know his initial motives. Was he aware that the early geologists of the day saw the Southern Uplands as a problem? For the strata of this terrain—over 50 miles across—seemed to them to be made of a single monstrously and implausibly thick unit. The graptolites in them had been noted, true, but as successive seams seemed to have the same graptolites, they were regarded as organisms that—unlike other fossils—did not appear to change through geological time. Seemingly stuck in a kind of time-warp throughout their existence, they were regarded as hopeless time-markers.

Lapworth looked closer. I would hazard that at first it was simply the strangeness and elegance of these fossils that attracted him, the Scottish examples (unlike the black-and-gold Welsh ones) being preserved in brilliant white against their dark rock background. Amid the landscapes of Walter Scott, their air of Celtic mystery must have been irresistible. He made a plan of campaign that would eventually encompass 300 square miles.

Within this vast terrain, he sought and then minutely analysed the black shale layers that occasionally interrupted the sea of sandstone. He found that, within any one mudstone unit, the graptolites were not the same, but, every metre or two, changed from one set of species (that he was simultaneously recognizing for the first time and naming) to another; in all, he recognized ten different assemblages. Some were mostly made of straight graptolites with sawtooth projections on either side; others included distinctive V- and Y-shaped forms; yet others were dominated by curved graptolites, like ornate fishhooks. So the graptolites were not frozen in time, but changed through time—and rapidly at that.

A few miles across country, though, there would be another seam of black shale within those endless sandstones, and farther away another one still. When Lapworth checked them out, they showed the same successive assemblages of graptolites. Had these fossils re-appeared in the same order, again and again, in those ancient Scottish seas, Groundhog Day style? The French-born scientist Joachim Barrande, then independently discovering graptolites in Bohemia, thought so. Lapworth disagreed, and said that the very strata had been folded over and over many times by massive earth movements, so that the one shale layer, with its rapidly evolved successions of graptolites, reappeared at the surface again and again and again. Lapworth was, in essence, correct. At one stroke, he had solved the riddle of the Southern Uplands in reducing its strata to a more modest thickness, and established the graptolites as extraordinarily valuable time-markers.

The graptolites continue to be biological chronometers, helping geologists unravel the complexities of mountain ranges and the vagaries of vanished oceans and ancient climate systems. Now, though, Lapworth's system of successive assemblages, or fossil zones, is elaborated. Today, instead of ten, there are now over sixty such time slices, each less than a million years long on average—which is good going when one is looking back to a world that is getting on for half a billion years old.

Precise time needs precise control on biology, on the recognition of individual species—so, instead of the handful of graptolite species recognized in Lapworth's day, there are now several thousand worldwide (and 697 in Britain alone, at the last count). It needs, too, a very good idea of when they originated and became extinct, which can only be established by patiently hammering through rock successions and logging the presence or absence of each species, stratum by stratum, and then, by writing the thick monographs and range charts that form the standard references of the palaeontologist's art. These will be well-thumbed as one searches the literature to identify, say, the pebble specimen, by now fully uncovered after perhaps a morning's hard labour with the needle.

If you are fortunate, that single specimen will be enough to establish the age of the pebble, to within a single graptolite zone—that is to say, within a million years or less. Some species have the happy combination of being highly distinctive *and* abundant *and* short-lived, bursting onto the scene, briefly— for a few hundred thousand years, perhaps—making hay just about everywhere, then disappearing (rather like the human species, perhaps, if the more gloomy prognostications of our future are correct). Or it may be a species that is longer-lived, that ranges through perhaps three fossil zones. Well, that's not too bad: better than if it ranges through six zones, as some tediously persistent species do. Or, after all that work, you may not have a complete specimen; rather, it is a fragment that may be any one of a dozen species, as the taxonomically crucial parts of the graptolite are missing. Well, there may be another gold-and-black fleck on the other side of the pebble that can be excavated. If one invests another hour or two digging away at that, maybe there'll be better hunting there. Or, you can call in your colleagues who specialize in acritarchs and chitinozoa to set to work with the strong acids and fine sieves and their own heavy stacks of monographs.

With the menagerie in our single pebble, there will be a way of wresting from it its own particular secret of time. There may be fits and starts on the way (acritarchs, for instance, are so tough that they can be eroded from one rock to turn up in another, as rogue time-markers). Nevertheless, the fossils will allow us to put that menagerie in its correct place within the evolving dynasties of life on this planet. Marvellous—but one must recall that the fossils are not the whole of the life of that time. They are but an echo; around them there was a much richer life, and we must have awareness of those potential riches, and of how we might unlock them, if we are to have any chance of deciphering their secrets in turn.

Ghosts in absentia

THE SOFT MACHINES

One of the books that changed my perception of the world is *The Open Sea, Part 1*, by the marine biologist Sir Alister Hardy. He had set out to write one book about the sea, but found that there was so much to say about the world of the plankton that it took up a whole book (he then had to write another book about everything else). It's now more than half a century old, and yet this hidden world remains marvellously evoked by his words, and by the antique black and white photographs and line drawings.

Coming to this as a palaeontologist, it was eye-opening. I was aware that in the strata, one normally only finds the remains of those forms of life that had some hard parts to fossilize. Bones, teeth, shells—and in the case of the acritarchs, chitinozoa and graptolites, their tough organic casings and homes. I knew that there had been other soft-bodied things out there of course, but alas these don't register often enough on the radar of the geologically programmed. So the sheer variety and exuberance of this world, revealed in those pages, took me by surprise. The remains of some of this life, within the pebble, lie somewhere within the amorphous black carbon that gives this object its dark colour, and in some of the subtle chemical signals of the rock itself. Parts of the hidden Silurian sea are beginning to be decoded from this unpromising material, and the stories emerging—fragmentary, ambiguous, tantalizing—sometimes have surprising uses.

Tow a fine-mesh net behind a ship for a few minutes, as Hardy did as a working scientist, and then examine its contents with a microscope, and a small fraction of this world is revealed—enough to reveal its almost boundless diversity. There are microscopic plants, the base of the food chain: the diatoms, for instance, single-celled algae with a silica skeleton that looks like a tiny ornate hatbox; the coccolithophores, even smaller algae with a bizarre calcium carbonate skeleton made of overlapping shield-like discs, and the dinoflagellates, too. Abundant and enormously important today, they had not yet appeared in the Silurian (we know that, as they are skeleton builders), so their ecological role must have be taken by other organisms, of which we surmise the acritarchs to have played their part, alongside unknown other green algae.

But it is the microscopic animals that catch the eye. There are single-celled ones, such as the radiolaria: amoeba-like organisms that secrete a complex silica skeleton. These were present, we know, in the Silurian and even earlier, because they can be seen in some rocks such as limestones, and cherts, the remains of deep-sea silica oozes. One suspects that there are the remains of a few of these in the pebble, but it is (so far) impossible to separate them from the pebble rock, as the weak acids that can liberate them from, say, limestones, will have no effect on the pebble, while stronger acids will dissolve both rock and radiolaria. One thus awaits someone to develop an ingenious technique to make them show up, and thus reveal another facet of this Silurian sea.

Then there are the creatures that are now simply carbon atoms, decayed, recombined into the tissues of other creatures (including those of the chitinozoan animals and the graptolites), turned to marine snow, then processed once more through the patient yet thorough microbial mats of that stagnant sea floor. What would there have been there? Hardy's antique wonderland shows a parade of tiny crustacea, with exoskeletons, yes, but so thin and delicate that none survive the recycling mill of the sea: the euphasiids (krill) and the naupliids, the copepods and the larvae of bottom-dwelling crustaceans. Then there are those creatures that are entirely gelatinous or diaphanous, the comb jellies and arrow worms and salps; and the jellyfish, growing to giant size by comparison, perhaps the biggest creatures in those seas. There are those swimming molluscs, the delicately shelled pteropods, or sea butterflies that are, alas, among the creatures hardest hit by the man-made acidification of the sea.

What place then would the graptolites and the chitinozoan creatures have had amongst this multitude of life? Graptolites, in these rocks, are the most obvious fossils by far, to the extent that the strata bearing them are called graptolite shales. The reconstructions show them as dominant in the open seas—yet did they really constitute the majority top predator? Or were they perhaps just 10 per cent or so of this niche? Or did they make up less than a hundredth part of the larger plankton, and only loom large in the fossil record because their resistant living quarters survived into the strata, and the delicate carapaces and tentacles of their more successful competitors didn't?

I would guess at something nearer the latter, not least because if the graptolites had been the main consumer of the primary producers, of the single-celled planktonic algae and such, then one would expect their remains to practically make up some of these deep-sea rocks: they should form a kind of graptolite coal (as the pteropods give rise to pteropod oozes today, sediments practically wholly made up of their remains)—and graptolites never do make up such a rock. Mind you, that reflects an ignorance of what happens to the graptolites after the death of their colony. Was it the fate of most or all of them to fall to the sea floor after death and thus be preserved, given the toughness and hence presumed inedibility of their organic-walled homes? That would suggest that the preserved graptolites that we see are mostly all there were, and hence that they were rarities of the Silurian plankton. Or were there scavengers out there capable of recycling even the seemingly inedible graptolite condominiums within the water column, so that only a small fraction of them fell to the sea floor to make it into the fossil record? That would allow one to infer a much greater standing population of these colonies.

It is one of the many uncertainties involved in this business of trying to peer back into the past. There are others, also predicated on comparisons with the nature of the plankton today. For instance, animal plankton today do not simply float or hover aimlessly in the surface waters, waiting for lunch to come by. Every day, as was evident even in Hardy's day, most of them take part in co-ordinated vertical migrations, swimming distances of as much as 100 m up to the surface waters at night, and then down into deeper waters during the day. For creatures only a few millimetres in size, these are enormous distances. Why make the journey? One idea is that the sunlit surface waters are more dangerous during the day, when predators can see the plankton more easily. Would that necessity still hold for the Silurian, when fish and whales and cuttlefish and squid had either not evolved or, if they had, they clung to shore, not yet invading the open oceans. One cannot say—yet.

A few predators of larger ambition (for those days) did lurk, though, among plankton. Occasionally, graptolites have been found that have been neatly and carefully folded into paperclip-like shapes, or crumpled more roughly into a ball. They were evidently predated. But what did the eating? The predator remains anonymous, though its eating habits are evident. Besides a talent for origami, its habits were neat: there was no tearing or shredding or biting or chewing. Such impeccable table manners were unlikely to have been a consolation for the poor colony, whose soft parts were probably digested whole.

Our pebble is highly unlikely to include a predated colony, for these examples are rare. But it might, just, include evidence that would have some bearing on the possibilities for daily vertical migration of the Silurian plankton. For today's oceans, mostly oxygenated throughout, don't offer a substantial barrier to the movement of plankton up or down. The partly anoxic oceans of the Silurian, though, might have presented a more serious hindrance to the vertical movement of plankton. The scale of that hindrance would have depended on whether the oxygen starvation extended most of the way through the water column, into the sunlit surface layers (thus compressing the living space for the plankton) or was mostly confined to the bottom waters, just above the sea floor.

One member of the plankton might have left clues as to which of these two alternatives is correct. There is today a type of green (i.e. photosynthetic) bacterium that also, remarkably, has to live in anoxic waters—it is a sulphur-metabolizing bacterium that does not tolerate free oxygen (technically, it's termed an obligate anaerobe). It produces a unique chemical that has the impressive name of isorenieratane. This is its photosynthetic pigment, a complex long-chain organic compound that has remarkable staying powers itself: it can be extracted, by extremely delicate chemical analysis, from rocks that are hundreds of millions of years old. It has not yet been found in the Welsh strata that yielded our pebble. This, though, may be a matter of time, as isorenieratane has been isolated from some North African strata of early Silurian age, showing that there, at least, deep-water anoxia persisted upwards to very nearly the surface of the sea.

Perhaps such inhospitable waters were a good thing for the graptolites. Being such byzantine and unlikely creatures, they may have had a tough time of it in normal marine waters, where the less sophisticated but more mobile (and to us, invisible) crustaceans and arrow-worms and jellyfish and such might have thoroughly outcompeted them. Certainly, fossilized graptolites are rare in strata that represent shallow, near-shore seas. It has been seriously suggested that they were adapted to low-oxygen conditions of the deep seas, where other plankton could not easily venture; these, of course, could not have been entirely oxygen-free, so these home-building colonies might have had to tread a fine line between suffocation and food security. The history of the planktonic graptolites is one of booms and busts (the busts often coinciding with episodes of oxygenation of the seas) so there may be something

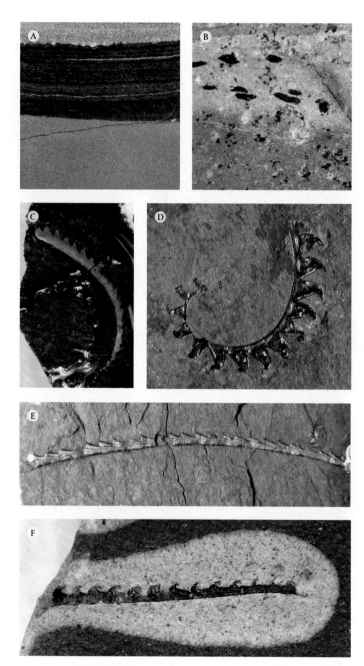

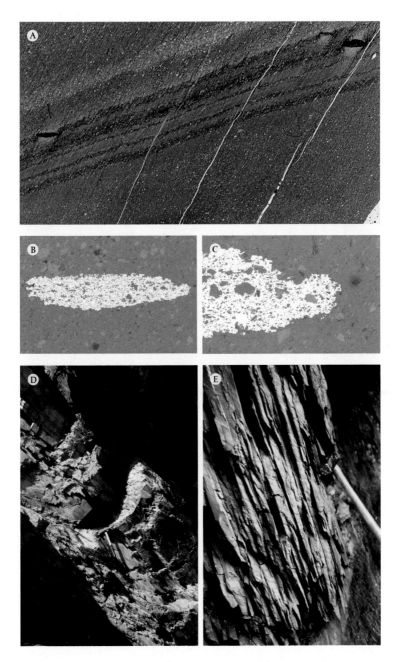

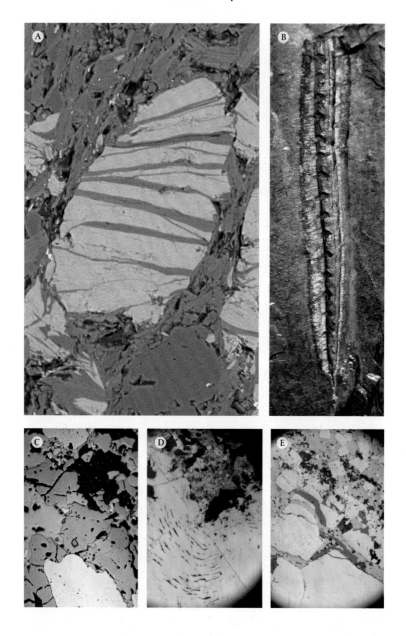

in the idea of them precariously clinging to an ecological tightrope—a tightrope from which they finally tumbled in mid-Devonian times, some 50 million years after our pebble stratum was being deposited, never to return.

There are yet other potential dimensions out there, that study of the modern plankton reveals. Were the graptolite colonies—and the chitinozoan-creatures, for that matter—green? Many sea-creatures today are part animal and part carefully tended garden, in that they include within their bodies unicellular green algae. The algae obtain shelter from this arrangement, and waste products from the animal as nutrients; the animals obtain an extra supply of oxygen, and also a part share of the carbohydrates that the algae manufacture. The bottom-dwelling reef-forming corals also engage in such inter-kingdom alliances. The symbiotic arrangement is so crucial to them that, if conditions become too uncomfortable for the algae (if the waters become too warm, for example), and they part company, that subsequently 'bleaches' the coral (it loses its pigment) which then usually dies.

It is still not possible to establish whether or not the graptolites and other Silurian plankton formed symbiotic associations with algae. However, that does not mean that that is an intractable question. It is not, as one politician of notoriously inimitable eloquence might have put it, an unknowable unknown. Such algae affect a host animal so much today (mostly for the better) that some clues may be sought, perhaps in the chemistry or molecular structure of finely preserved graptolites. Or perhaps (trickier, this), in modelling of the growth of the graptolite—in particular to see how it managed to divert so much of its resources to building its large and robust and highly energetically costly living quarters. Did symbiotic algae make this possible? At the moment, this idea is surmise (wild surmise, some would say) and not (quite) worth a light: it will become science if a way is found to put forward some evidence for or against it, and that is the exciting—and frustrating—part of this entire pursuit.

Graptolites and chitinozoan creatures may or may not have been green—but they must have fallen ill from time to time, or even succumbed to pandemics (as many species of amphibian, today, are falling victim to fungal infections, for reasons unknown but probably exacerbated by factors such as climate warming and industrial pollution). This kind of thing is extraordinarily difficult to gauge from the fossil record, as many infections do not leave a trace. Some parasites, though, can leave their mark more clearly. Examples of graptolites have been found that have large blisters on them, where parasites had obviously attached themselves, and the graptolite zooids had then done their level best to 'plaster' them over—literally to wall them in. The fightback seems to have been successful (the graptolite colonies carried on living, for a while anyway) but only up to a point. Colony building often went markedly askew after parasitization, this being understandable—could you carry on building that perfect patio while a burglar clung to your back, hitting you on the head with a blackjack and demanding your credit card?

Whatever part simple illness played in the lives and deaths of the pebble plankton, it is astonishing that only in the past few years have scientists discovered just how rich in bacteria and viruses the modern oceans are. In the days when these were mainly studied by making cultures in Petri dishes and by microscopic examinations, some few thousand microbe species had been painstakingly isolated and named. Then came the technological breakthrough of automated gene sequencing, and *millions* of bacterial species emerged from just a few litres of standard seawater, instantly revolutionizing our picture (if not altogether our understanding) of microbial diversity in the oceans. And if the bacteria are now seen as almost infinitely diverse, then for each microbe there are ten viruses in the oceans, which constantly prey on those microbes, and on other organisms too. Between them, these organisms govern the biology and chemistry of the oceans, although many of the particulars of that governance are still obscure to us.

They, for sure, would have been at the heart of the governance of the marine ecology of our pebble stuff too, but obtaining scientific evidence of that central reality is probably something that lies decades (at least) into the future. And, as science is the art of the possible, one has to go where one can, to look at the grand, if shadowy patterns and histories that were unfolding themselves as life evolved on the Silurian Earth.

CHEMICAL MESSAGES

What, for instance, was the sum total of life, of living organisms, in the oceans that stretched out from that particular spot of the ancient Welsh sea floor, a few square centimetres across, that much later became the pebble? This is a huge, almost abstract question. Nevertheless, potential clues are there, even within the pebble material itself. Their meaning is still ambiguous, but the evidence is now technically easy to come by, so such evidence is currently being amassed and studied by scientists around the world, in the hope of illuminating another side of the Silurian world. It is science as the art of the possible, here at work, and it would here investigate the nature of the dark, carbon-rich component of the pebble.

Carbon comes in two main stable forms, a lighter isotope, ^{12}C, with six neutrons and six protons in its nucleus, and a heavier isotope, ^{13}C, with seven neutrons and six protons. These have the same chemical properties, but because of their difference in mass, are often segregated in chemical or biological reactions. Photosynthetic plankton, for instance, absorb both isotopes—but show a distinct partiality for ^{12}C, and will preferentially absorb more of this light isotope in building their bodies—and consequently this will leave a little more of the heavier ^{13}C in the environment around it.

Now, when the good times roll, large amounts of plankton will live and die. If they then fall to the sea floor and are buried in sea floor sediments, taking their excess of light carbon with them, the oceans above will be enriched in the heavy carbon isotope. Subsequent generations of plankton still segregate out the lighter isotope but, because there is less of that in the environment overall, these succeeding plankton will contain more heavy carbon than their predecessors. And so when these in turn fall to the sea floor and are buried, their particular stratum will have a greater proportion of light carbon atoms than the ocean water above—but a smaller proportion than in the underlying layer of sea floor mud.

It is a grand atomic reshuffling exercise that can take place on a truly global scale. If life thrives for a good long while, and is then buried, then the whole carbon isotope balance of the world can change. This kind of change can now be detected, routinely, by modern atom-counting machines. It just takes a few grams of carbon-bearing mudrock, so a small part of the pebble will suffice. That needs to be crushed, flash-heated and converted into a plasma; the ions in the plasma are then whirled around a curved track, within a magnetic field. The lighter ions are deflected less by that field than are the heavier ones, and so the different atomic weights separate out before the ions slam into carefully positioned detectors, that count the number of impacts of each type (that is, each weight category) of ion.

This unambiguously and precisely gives the ratio of light to heavy carbon in the pebble sample. Now as to what that *means*—that requires a little more thought. Such a number is really only significant in context: that is, relative to the ratios of the carbon isotopes in the strata above and below. Thus, one would have to work out exactly which level in the stratal succession of Wales the pebble came from—that is, how old it was. If one took all the evidence from the acritarchs, and chitinozoans and graptolites—and carbon isotopes—from a single pebble, there would be a fighting chance of doing just that, albeit with the sacrifice of most or all of the pebble.

Ideally, the pattern of carbon isotopes through a stratal succession should show a pattern that will reflect some major environmental events (such as the flourishing or otherwise of life) and that will be repeated in other stratal successions of the same age elsewhere in the world. That is not always the case, but it is becoming clear that some episodes of the geological past were characterized by major changes in global carbon isotope composition; these are termed isotope 'excursions' and seem to reflect major environmental perturbations. There was one such excursion just before the beginning of the Silurian Period, for instance, associated with that brief but severe pulse of glaciation that devastated so much marine life. During this interval, the carbon in strata around the world becomes progressively and markedly enriched in ^{13}C. Does this reflect a flourishing of life and its burial, as described above? Perhaps not, for there are other ways of shuffling carbon isotopes.

Limestones, for instance, typically contain significantly more of the heavier carbon isotope than does the organic material in mudrocks. Therefore, if you take a lot of limestone and dissolve it into the world's oceans—as is possible when sea level falls and coral reefs are exposed and eroded, for instance—then those oceans, and the strata that form beneath them, will contain more heavy carbon.

Looking through the prism of different types of carbon atom therefore gives a sideways and ambiguous perspective on the world of the past. However, this picture is also, potentially, immensely insightful. As more and more data is generated with the help of those marvellous mass spectrometers, then the picture of, and reasons behind, these enormous global atomic reshuffles will—one hopes—become clearer.

And there is more, perhaps, that could be gleaned from the carbon patterns, even from within the pebble itself. For in today's oceans there are differences in the carbon isotope ratio between the predator and the prey. The catchphrase here is 'you are what you eat—plus one part per mil' which means that what is left inside you of the food that you eat is enriched relative to it, by about one part in a thousand, in the heavy carbon isotope. One might expect similar relationships between the graptolite (as the predator) and what one might regard as prey: the acritarchs, perhaps. As the atom counting machines are developed and modified to analyse ever-smaller amounts of material ever more quickly and cheaply, it is becoming possible in practice to separate out tiny fragments of different fossil from a rock and to analyse these. Watch this space.

In modern oceans, too, there are isotopic differences between, say, shallow and deep parts of the water column. If the same held true for Silurian oceans, could these differences be exploited? Well, today one of the enduring problems of graptolite ecology is whether different types of graptolites lived at different depths, with some perhaps at the very surface of the ocean and others much deeper down. Such a possibility has been long suspected—for instance, some graptolites are very robustly built (perhaps to withstand strong wave action), while others are much more slender and delicate. The trouble is that once a graptolite has landed on the sea floor, there has been no clear evidence left to tell quite how high up in the water it lived. Perhaps if different types of graptolite can be shown to have different proportions of carbon isotopes too, then that might help shed light on that particular riddle. Perhaps. Isotopes in geology can provide marvellous pictures of the past—but also mirages. They need to be approached with care. The trick, as ever, is to ask the right kind of question, tackle it by means of appropriate analysis, and to look at the answer one receives with due scepticism. It's work in progress.

Still, there might be a parting present from the fossilized life forms, recognizable and unrecognizable, of the pebble. They may be able to measure the time, in years, since the time they lived, courtesy of their friendly relationship with some very, very rare elements. Fossils, of course, are good at telling the time. Practically indispensible, in fact. But the time they tell is relative time. Using them, one can tell whether one fossil-bearing rock is older than, younger than, or the same age as another fossil-bearing rock, often to a very fine degree. But to tell time in numbers of years has been largely the province of the radioactivity clock, and of the minerals that contain radioactive elements that break down into other elements at a known rate, as was discussed in Chapter 3. Thus, to get the age of a stratum of Welsh mudrock, we need to see if, within it, there is, say, a layer of volcanic ash with those wonderfully precisely dateable zircon crystals. Alas, such layers are rare generally in the Welsh Silurian, and one would certainly not expect one in the pebble, for the ash layers are extremely weak, and the rock breaks easily along them.

But, there is another, much more exotic element out there that can be exploited. This element is rhenium, and it is much rarer than gold, being measured in parts per billion. However, it has two characters that make it stand out as a chronometer and, as such, is the new kid on the block when it comes to dating sedimentary rocks. First, it is radioactive, decaying by the loss of an electron into the equally rare osmium (which, as a metal, has the distinction of being the densest known, at a staggering 22.6 grams per cubic centimetre). Secondly, it has an affinity—as does osmium—for iron, for sulphur, and for organic matter. Thus, in black shales (and, indeed in oil) there is just about enough to analyse for these two elements, and to essay a date from them.

It is a long and complex and expensive procedure, and not for the faint-hearted (or impecunious). One also needs to obtain not one analysis, but several, for obtaining the date depends on obtaining what is called an isochron: that is, several pairs of rhenium/osmium ratios with a consistent relationship between them (so although it is inherently a long process, it has a built-in reliability check). The poor pebble, thus, would have to be sacrificed once more, cut into several pieces, one for each analysis (there should be *just* about enough material) and put through a minutely careful procedure to isolate the tiny quantities of these elements, and to measure them on the mass spectrometer.

To obtain a reliable radiometric date from a sedimentary rock, without having to depend on the vagaries of finding or not finding layers of volcanic ash, is something akin to a holy grail for geologists. The rhenium/osmium clock is distinctly promising—it seems to work, and the few dates obtained so far have been consistent with other dates, although not yet as precise as the marvellous zircon geochronometer. Because it is very expensive and laborious, it is nowhere near to being a routine technique yet. In a few years, though, it might become one, thus opening up another vista of possibilities for charting the history of this planet.

Meanwhile, our humble pebble has yet more locked away in it of the world about it, while it is yet sediment on the sea floor. It is not yet time to bury that sediment, to enter the underground realm. There is the question of getting a fix on just *where* on Earth that little patch of sediment was.

Where on Earth?

In some ways the pebble is like one of the newer computer chips, tightly packed with more information than one could ever surmise from gazing on its smooth surface. That stored information can relate to any episode in the history of the pebble, and could be derived from nearby—a microbial mat growing on the exact spot on the sea floor where the pebble sediment accumulated, perhaps. But it could come from afar, such as a micrometeorite landing in the ocean and drifting slowly down to land on that very same spot (there are likely a few of those in the pebble, too). Some information is as pristine as the day it was written, in its own particular code, into the pebble fabric; some, on the other hand, has been almost completely overwritten, when yet further information was imprinted at some later point in time.

We might consider here some information that has most likely been all but erased by the pebble's tumultuous subsequent history—not that that should stop us trying to recover what we can of it. Nevertheless, when it was written into the fabric of the pebble, it provided a clear signal that travelled easily through some 4000 miles of solid rock, straight from the centre of the Earth. This signal gently nudged and guided certain of the flakes of sediment falling on to that sea floor. It made them line up, with almost military precision, to point polewards. They form a memory of latitude.

The Earth's magnetic field is a mysterious thing. What is magnetism? As a child, I used to push together the north poles of two toy magnets, and remember even now how frustratingly difficult it was to make them touch—or how tricky it was to prevent the north and south poles from locking together when I tried to keep them just a *tiny* bit apart. A few years later, I looked on, impressed but with incomprehension, as a physics teacher sprinkled iron filings around a magnet, to show how they lined up along the invisible lines of force. But what were these lines?—and why in that particular shape? It's still something I don't have any visceral understanding of, any more than I can explain how the innards of this computer produce letters on the screen when I press the keys. It is something that just is.

The ancients were equally puzzled (though with more reason to be). The magnetic properties of lodestone (the iron mineral magnetite) had been discovered by the Chinese over 2000 years ago, and they went on to invent the compass, by placing a spoon made of lodestone on a smooth board. It took a thousand more years for knowledge of this phenomenon to be appreciated in Europe, when it was observed that a lodestone needle pointed towards the pole star, the only fixed star in the sky, and that was naturally taken as the source of the magnet-ism (although there was an appealing counter-idea that there were mountains made of lodestone at the North Pole).

Then came the discovery of magnetic declination: that the compass needle did not point to true north, but a little way away, to magnetic north; and of magnetic inclination, that the needle also pointed downwards towards the pole (not at all at the equator, but vertically downwards at the pole—where compasses are useless for that reason). William Gilbert in 1600 first understood what this meant, by experimenting with how a sphere of lodestone (representing the Earth) affected a lodestone needle moved across its surface. He realized that it was not a star, nor magnetic mountains, but the whole Earth that acted as a magnet.

And now it is known (though this is something that I cannot truly fathom either) that the magnetic field is the product of the Earth's molten core (which has, though, a solid centre). It is the electric eddy currents in the slowly swirling molten iron-nickel outer core, allied to the Earth's spin around its axis, which produce the Earth's magnetic field. Looked at more closely, the Earth's magnetic field is a mobile, dynamic thing: the Earth's spin keeps it aligned roughly north–south but in detail the magnetic north and south poles are never stable but move across the Earth's surface by ten kilometres a year or more, to produce the constantly varying declination.

The field also flips every so often, north becoming south and south becoming north—lately it has been doing so every few hundred thousand years, though at some times it has remained locked into a single mode for tens of millions of years. In the Silurian, the magnetic field generally flipped frequently, though there was one 'quiet phase' near the middle of the period without any such flips.

For our pebble, one point of general significance is that this phenomenon was the guarantor of all of the life that became preserved within it. Without a magnetic field, there would be no shield to protect the Earth from the cosmic radiation of the solar wind, that would otherwise shatter any promising-looking complicated and delicate organic chemicals that might emerge on this planet, and for good measure strip away most of the atmosphere and oceans—as probably happened with Venus and Mars, both of which lack a magnetic field.

Much more specifically, it is the phenomenon that lines up certain sedimentary particles along its lines of force, to become tiny embedded compass needles in the sediment. These are particles of anything magnetic, mainly particles of iron oxide—goethite and magnetite as such, that form a small part of the silt and fine sand that the turbidity currents and nepheloid plumes have brought into the sea.

It is a signal that can be read from the rock, courtesy again of the ingenuity that humans have developed to probe the most subtle and unlikely properties of any earthly material. In the laboratory, one can measure whatever is left of the tiny amounts of magnetism imparted to the original sediment, within a small core drilled from the rock stratum. To do this *properly*, one needs to know how the rock lies in the strata, to be able to relate its present position to its ancient one. But, even the pebble might yield some information—as to what angle these little natural compass needles make relative to the Silurian sea floor.

For that is the key information—the higher the angle, the nearer to the north or south pole one is, and we can thus can work out the latitude. Longitude is much trickier—impossible using this technique, in fact, and largely educated guesswork even by other means. Nevertheless, latitude is pretty good. The pebble spot of the sea floor was at about 35 degrees south of the equator, which is at about the current latitude of the northern tip of New Zealand today, or the southern tip of Africa.

Since that time the pebble spot has moved, more or less steadily, northwards. Then, 50 or so million years later, in the Devonian and early Carboniferous periods, it was in the southern subtropical zone. Being then about 5 kilometres underground, it was largely indifferent to the weather at the surface (largely warm and semi-arid). Similarly, it crossed the equator about 310 million years ago. There was no fanfare, no ceremony, nor did the markedly different landscape above (a wet tropical swamp) make any difference that can be detected within the fabric of the pebble. And then, towards a quarter of a billion years ago, it was below a desert, fully arid now and baking hot, in a position akin to the Sahara today, in the northern arid subtropics.

From there it was a slow progression to its present position. As a patient crow flies, the distance is something over 10,000 kilometres (not counting that immeasurable movement from east to west), which it accomplished, almost entirely underground, in a little over 400 million years, at an average speed of more than 2 centimetres a year. And it is still going, of course.

Progress, though, was unsteady, for the fate of Avalonia was bound up with its larger neighbours to north and south. The frozen compass needles in the rocks betray the relations between them. To the north, what is now Scotland and North America were looming, and the Iapetus Ocean that once formed a barrier thousands of miles across, had closed to almost nothing. Southwards lay the huge continent Gondwana—of which what we now know as Africa formed the core—but it was over 1000 miles away, having been moving away for the best part of 100 million years. But the intervening ocean, the Rheic Ocean, having been as big as it was going to get, was beginning to close. In another 100 million years it in turn would shrink and vanish, as the very crust of the ocean slid into the mantle along subduction zones.

The frozen memory of these travels is a gift from the Earth's core. A surprising one, perhaps, for the world has changed much since the time of the pebble. The surface has changed, of course, with the shifting of continents, the growth and destruction of mountain belts and the evolution of new forms of life. But this is, in a very real sense, superficial. What has

FIGURE 5 The position of Avalonia in the Silurian.

really changed, hugely, is the centre of the Earth, its nickel-iron dominated core. This, today, is a sphere of molten metal 7000 kilometres across, inside which is an inner core, about 2400 kilometres across, which is of solid metal. This is not in a constant state. The core of the Earth is steadily and inexorably freezing, as the entire planet cools. Roughly a billion years ago (*very* roughly: estimates vary from over 3 billion to a mere half billion) there was no solid inner core. It has developed since, and in a billion years or so from now it will be all there is, for the entire core will have frozen—and hence the Earth's protective magnetic field will have disappeared. And there is nothing that we can do, Hollywood-style, about it.

The rate of this change is awesome. Let us say that the strata in the pebble represent something of the order of a century of sedimentation on the sea floor, near enough a human lifetime. In that time, assuming a rate of freezing similar to today's, of around 5000 tons a *second*, some 1.5 thousand billion tons of metal would have converted from liquid to solid—with the inner core thus grown in diameter by some 5 cm. In a certain sense the growing inner core is a planet evolving within a planet, for the seismologists and theorists who reconstruct it suggest that today it now has a kind of tectonics at its surface, with large sections of surface 'crust' rupturing and sliding over each other. And others have surmised that the inner core landscape is clothed in a kind of forest, of branching iron crystals growing out into the liquid iron above.

In the time of our pebble, of the Silurian, the structure and behaviour of the inner core may have been quite different. Being smaller, it was probably much more violently affected by vertical, convective movements of material within it, the motion being driven by the heat released as liquid turns to solid.

This represents a fundamental transformation of planetary character, albeit one enacted far beneath the surface. The Earth of the pebble's genesis was thus truly different.

But what more can we say about the position of the pebble spot on the sea floor? There is more that we can deduce from its tiny freight of particles and corpses, and how these might relate to those in strata elsewhere in and around that former landmass of Avalonia. The particles, for instance, though deposited far from land by turbidity currents, seem not to have been deposited on true ocean floor, in the way that, say, mud and sand from the Himalayas is carried far across the Indian Ocean floor. No trace of the kind of basalt that makes up real ocean crust has been found associated with the layers of strata from where the pebble came, as one traces them around the principality of Wales. Rather, this was a deep sea, but one that developed within the crust of the Avalonian continent, as its crust stretched and foundered, tugged by plate tectonic movements. Thus, this ancient Welsh sea was more akin to the North Sea, which is still subsiding and accumulating thick layers of sediment today, than to the Atlantic Ocean.

THE TOLERANCE OF ANIMALS

How else can one get a sense of position, relative to elsewhere? The fossils have yet another task to carry out. For, they are not only indicators of the passing of geological time. They are also indicators of geography, of where a place is on Earth. The use here runs counter to that of age determination, for the best fossil guides to geological time are those that not only lived fast and died young, but those that also colonized as widely as possible, so that they can turn up in—and enable the comparison of—strata from as many places across the Earth as possible.

Fossil guides to geographic place, though, need to be stick-in-the-muds and stay-at-home types—like the finches and giant tortoises of the Galapagos Islands, for example, with each island possessing its own distinctive species, or the dragons of Komodo. The distribution of each species is thus controlled partly by such conditions as temperature and food supply that are suitable for it, and partly by accident—where it arose—and partly by barriers that stop it migrating to other places where conditions may, in theory, have allowed it to thrive.

The pebble fossils are not, alas, like Darwin's finches, each bound to a particular island and hence characterizing (to a discerning eye) that island. Being plankton of the open sea, they could spread more widely, and the barriers that restrained them are more subtle. Nevertheless, today's plankton encounter their own barriers in the seas. They are sensitive to temperature, to nutrient levels, to suspended sediment levels. Particular species inhabit water masses with particular properties. The Silurian plankton of the pebble, likewise, could not and did not spread through the entire oceans of those days, but dwelt in those areas of the sea where conditions were right for them.

They ranged widely enough, though. Take for example that hard-won, delicately excavated graptolite from the pebble. You would certainly find its like from other strata of the same age in Wales. There is a good chance that you would find that very same species elsewhere around the shores of Avalonia, into what is now western Europe, say, and even further into central Europe, or across the narrowing ocean that still separated—just—Avalonia from what is now Scotland and north America.

But range more widely, towards the tropics of those times (a good example, ironically, is present-day Arctic Canada) and you will probably have less luck. The warm waters there were host to a different assemblage of plankton. It is unlikely that the pebble graptolite will be found there (but not impossible, for there were a few species in common, luckily for those who seek to throw time-lines across these former distances).

Ultimately, one would like to place the pebble graptolite in a much bigger context. That is, for each species (indeed, of any kind of animal and plant), to establish quite when it lived and where it lived; to understand its full distribution in time and space: in time, to optimize its use as a time-marker; in space, to help reconstruct the great biological provinces of the past and to determine the kinds of factors, such as climate, that controlled them. This is a huge task—the making of many inventories of life across many millions of years. And, aside from the cost in time and money and sweat (and tears, too, though generally it is a mainly bloodless exercise), there are some basic stumbling blocks.

How does one agree on what a species is, for example? In palaeontology, all one normally has is the flattened and crushed remains of some incomplete part of the carcass of what was once a living animal or plant. Biologists argue over what a species is even today, but a handy working definition is that it is a population capable of interbreeding and producing fertile offspring. Now, that is emphatically not a test that one can apply to fossils. Their breeding days are over. So, what one works with here are termed morphospecies—fossils that look nearly alike enough to be all called by the same name. How alike enough is alike enough? This is where the fur can begin to fly, for some scientists (the 'splitters') make very fine distinctions, and describe many fossil species differing in fine detail, while others (the 'lumpers') set much broader limits to what they consider to be a single species, allowing more for natural variation in size and shape (the way that different humans vary in these characters, for instance). Worse, quite a few species today are cryptic, which means that they either cannot be told apart by any physical observable feature (their differences perhaps being in their physiology or their behaviour), or the physical differences are so slight (and entirely within their soft parts) that they do not fossilize.

Faced with these (and other) problems, who would be a palaeontologist? A dog's life, surely. The astounding thing is that the science works as well as it does. There is often reasonable agreement among palaeontologists about what to call particular fossils (though admittedly that is *cautious* reasonable agreement), and that enables comparison—up to a point. It may sweep a lot of taxonomic problems under the carpet, but then, every now and again, that carpet will be lifted up, and what is underneath can be closely scrutinized. There are other species, of course, that are obscure, poorly defined, controversial—well, in making global analyses one has to work with the data as best as one can, warts and all.

The general picture can be clear enough to appear even through such noisy, imperfect, partly subjective data. Some graptolite and chitinozoan assemblages, laboriously collated from around the world for particular 'time slices', show abrupt changes from very diverse assemblages, with many species, to much less diverse ones, as one goes from low to high latitudes. Today, similar changes in modern plankton are associated with the position where the cold, Arctic and Antarctic currents meet the warmer temperate seas. It seems that such a 'polar front' effect may, too, have affected the plankton of those long-vanished oceans.

And then there are the quirks. Odd species that do odd things. One graptolite species, say, that, when its occurrence is plotted on reconstructions of the Silurian globe, seems to stretch from tropics across into temperate climes and then carries on towards the poles—an astoundingly broad range. Was it simply a species more tolerant of extremes of heat and cold than any form of modern plankton? Or was it one species that changed its position in the water so as to try to keep conditions around it the same: that lived near the water's surface at high latitudes, but that descended to cooler, deeper waters around the tropics. Or was it not one species but several, that look alike, lived in more or less the same time-slice, but differed in their physiological properties? As this same species appears to have 'broad' and 'narrow' varieties in some different parts of the world, the last of these possibilities may be the most reasonable interpretation—for now.

It just shows that when you take your handful of fossils from the pebble, and start to dig into their characters, their relations and their histories, the trail of clues that you find can lead you half way around the world. Admittedly, there are typically a whole string of wild goose chases along the way, but that is the charm of the whole exercise—one, moreover, that shows just how much we have to learn, how near we are to the beginnings of a very young science.

A WALK IN THE COUNTRYSIDE

One can look more locally, too. For a detailed sense of the geography of times past, one must sometimes step outside of the confines of a pebble, and take stock of its context—of *all* of its context, or, at least, of as much as one can humanly get at. The pebble is a tiny part of a Silurian sea floor that extended for many miles in every direction. This sea floor had its own particular geography: parts were shallower, parts were deeper, some expanses were flat, some regions formed steep scarps. This geography had an influence—indeed, it had many influences: on where oxygenation of the water finished and anoxia started, and hence on what kind of life could exist; on the pattern of tides and waves at the coast; and, perhaps of most importance to the production of future rock, on determining the course of the sediment-laden turbidity currents, as they swept in from shallow water.

How does one make an exploration across an ancient sea floor? Why, one can simply walk across it. There is no need for pressure suits, or oxygen cylinders, or bathyscaphes. Just a good pair of boots, trousers that can withstand brambles, a rain-proof cagoule and a sunhat—for the weather in these parts can be pleasingly various. In walking across the countryside, one is walking across the strata laid down on that sea floor, and one has all the necessary evidence to transport oneself back across its original contours. If the pebble is the microcosm, here we have the macrocosm. It is called field geology.

There are adjustments to be made, mind you. One hardly ever walks across a single stratal surface that represents a sea floor at one instant in time. One walks across and past thicknesses of strata, that represent segments of the history of that sea floor. One does not therefore seek a single tableau, a panorama of the past. One is after a moving picture of a sea floor, simultaneously evolving across space and through millions of years of time. There are all the practical problems.

Wales is a green and pleasant land, after all, and most of the rock strata are hidden by soil and vegetation, and also by the debris left by the last ice age. So one has to go in search of the rock, along streams and gorges (hence the bramble-proof trousers) and on hillside crags, and in farmyard quarries too. The rock has a long history, so it has been crumpled, dislocated by earth movements: one then has a four-dimensional jigsaw to put back together.

Field geology is the ultimate forensic science, the art of the possible, where one combines as much evidence as one can get hold of, with as much ingenuity in analysing it as one can muster—and also with a keen sense of the limitations of one's deductions. This is an *idea* of a sea floor that we are building in our heads, not its bones that we can precisely reconstruct in some grand museum. It is an idea that can be tested (to destruction, sometimes) by gathering new evidence; it can be revised, honed, built upon—but it is an idea neverthe-less, a working model that will evolve as our understanding evolves.

None the less, our pebble came out of the cliffs of central west Wales. It forms part of a much larger unit of strata—a longer stretch of history of the sea floor. They are strata that have a long history of study, too. As they are superbly exposed in the cliffs around the large seaside town of Aberystwyth, they have attracted much attention down the years. Indeed, this is one of the places where such rhythmically striped rocks were first interpreted as having been laid down by that undersea phenomenon, turbidity currents—quite a breakthrough in geology. As the early researchers studied the cliffs from north to south, they saw that the layers became thinner the farther north one went, and that the sediment grains making them up became finer. Thus came, entirely reasonably, the deduction of the structure of this turbidite fan, as a mass of sediment that progressively thins and becomes finer-grained along the course of the steadily weakening turbidity currents.

It was a standard model for many years—indeed, one with which strata all round the world were compared. And it is right—but wrong. The early researchers kept their studies to a north-south line along the cliffs, because that is where the rocks are most easily seen. To the west is the sea—so there one cannot go, or at least not without a multi-million pound budget and a submarine. And to the east lie the green hills of Wales, with the crumpled rocks that poke out just here and there.

To look at those rocks, and draw a picture of a sea floor out of them, one simply has to walk the hills and valleys, examining, describing, measuring those strata, and interpreting them—as we have interpreted the pebble in these pages—in terms of what took place on the sea floor, and when. The 'what' can tell, say, of how powerful the ceaseless turbidity currents were, and how much sediment they were carrying and depositing. For the 'when'—without which all the strata would be an unintelligible rock soup—there are the fossils, and in particular the graptolites, in search of which one has to scour the rock exposures with a keen eye—these creatures are small! With these fossils, one can divide the strata into units, each representing perhaps half a million years of history. Thus, one can track the time, as well as the rocks, across the countryside.

When this study was made, a few years ago, the sea floor changed shape, and dramatically, at that. Or rather, its idea in human minds changed shape, and came closer—one hopes and trusts—to that real sea floor, the one that existed in Silurian times. For here the strata did not so much change from south to north—although that trend was still visible. They changed from west to east—and how! Towards the east, the strata (not insubstantial in those sea cliffs, of course) became immensely thicker, going from a few hundred metres thick at the coast to over *two kilometres* thick just a few miles inland. The individual strata, the units of sediment deposited by individual turbidity currents became much thicker too, some attaining two metres and more. Then, a little farther east, these gargantuan strata just disappeared. All that was in their place (and again, the little graptolites kept steady one's sense of fossil time) were some thin mudrocks. What was going on?

The celebrated cliff exposures turned out to be just the thin edge of a monstrous wedge of sand and mud that had been pouring into the Silurian sea of Wales. It had been coming from the south, true—those early observations were not wrong. But most of the sediment had, quite naturally, been piling up in the deepest part of the sea which (in present-day geography) is now inland. As the sediment poured in, the sea floor subsided, and so more sand and mud flowed in, pushing the sea floor down further, and so on—this was positive feedback in action. And the sea floor did not simply sag down in a fold. It opened like a trapdoor, with the hinge in the west and a massive dislocation—a geological fault—in the east where one mass of strata simply broke away and slid vertically downwards in the crust. This left an undersea cliff on the other side, which confined and directed the onrushing turbidity currents along its foot, preventing them from passing farther eastwards. A half-graben, it's called—a common enough feature in landscapes past and present, but this one is a particularly fine example. It opened, sagged downwards, filled with sediment and then seized up in little more than a million years.

FIGURE 6 A half-graben.

It's a fine panorama, a seafloorscape to rival any of the landscapes in the hills around. A fitting context for the pebble sediment while it was still at the surface. It is now time, though, to depart from the brief repose of the pebble sediment at the surface of this sea floor, and enter the much longer time of its imprisonment underground. Prison, in this case, was a place of reform.

Gold!

OF MICROBES AND METAL

This is the beginning of the long goodbye to the surface realm. The flakes and grains of the pebble material are now in utter darkness (except perhaps for occasional flickers of phosphorescence from some of that microbial life), at the bottom of that deep, stagnant sea. The strata that we see in the pebble are a few centimetres thick. But now, of course, they are made of good, hard, respectable, tightly compressed rock. Back then, they made a layer of mud—waterlogged, sticky, slimy, and very likely evil-smelling mud—a quarter of a metre thick or more, that formed part of a layer on the sea floor that extended for tens of kilometres in every direction. Let us catch it at *just* this point in time, before it became buried by further influxes of sediment from those endless turbidity currents.

The mud was full of life, particularly at the surface, most of which will have been occupied by those infinitely complex microscopic city-states that are microbial mats. But even below that, in the buried mud itself, there will have been considerable activity. In fact, as microbes are extremely good at clinging to life in all kinds of conditions, that activity was to carry on for quite some time yet. Those indefatigable microbes, though, still had to earn their keep. One way of doing that was by making use of the soft tissues of the fallen plankton, that were dismantled and recycled in the process that we call decay. Even in these anoxic conditions, where decay was slow, the magnificent, complex molecular architecture of body tissues was beginning to degrade, to transform into smaller, simpler molecules, leaving just the considerable inedible remnants that are the cases of the acritarchs and the chitinozoa, and the living quarters of the graptolites, upon which the microbes did not seem to manage to get much of a foothold (so to speak), even though they had decades and centuries in which to make the attempt.

It is one thing to be occupied in this microscopic breaker's yard, amid the wreckage of proteins, fats, and carbohydrates. But there are other tasks, one of which is the seeking of energy. And this led to a surprising—but very beautiful (to human eyes) by-product.

We, of course, use that rocket fuel (literally) of oxygen, which is the most effective means to burn the fuel of our food. In the absence of oxygen, one uses the next best thing. In seawater, the next best thing is the sulphate (SO_4^{2-}) ion, which is plentiful. Indeed, if seawater is evaporated, one of the main products after sodium chloride (rock salt) is calcium sulphate, otherwise known as gypsum, the stuff that one makes plaster of Paris from. If a microbe strips the oxygen atoms from sulphate, the sulphur ends up by itself, not as free sulphur, but as the sulphide (S^{2-}) ion. And where that meets up with the dissolved iron ions, they combine to make iron sulphide, FeS_2– which is pyrite, sometimes called iron pyrites, more popularly known as fool's gold, because of its beautiful golden yellow colour.

Has it really fooled many prospectors? Naive greenhorns—or hopeful and gullible investors—may fall for it the first time, perhaps. Pyrite is a common mineral, and anyone who seriously looks at rocks for a living will encounter it very soon in their careers. It is not anywhere near as heavy as gold, is brittle rather than malleable, and tarnishes easily. Gold deposits normally form deep underground. Pyrite can form in the subterranean depths also, to be sure—but it can also form much nearer the surface— just under your feet, for instance, as you take a walk along a beach.

In the deep stagnant water, above the sea floor, something was probably happening directly above the little spot where the pebble was forming. Tiny golden crystals of pyrite are simply appearing out of, and suspended in, the dark water. They are cubes, or variations of cubes—with corners or edges bevelled off, say. And somehow, as they formed, they arranged themselves into a three-dimensional array, one of great geometrical elegance: a sphere made up of many hundreds of individual micro-crystals. It looks a little like a raspberry, this compound microscopic structure, and its name reflects that. It is a pyrite framboid (from *framboise*, French for raspberry). When it got so big that the water could no longer support it, at about nine thousandths of a millimetre in diameter, it began to sink through the seawater to fall into the mud below. It was snowing framboids, and some of that snow will have fallen, for sure, into the tiny patch of mud that was to become our pebble. Within the mud itself, other framboids will have been forming, by the same process and from the same materials. This time, though, as they are supported within the thick mud, they could grow larger, often to a tenth of a millimetre across or more.

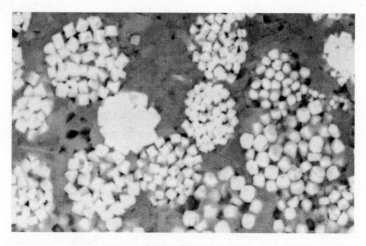

FIGURE 7 Pyrite framboids, highly magnified.

How are these remarkable structures formed? Are framboids essentially fossilized bacterial colonies? Scientists have suggested that. Well, that doesn't quite seem to be the case, not least because framboids have been grown artificially in a sterile environment in the laboratory, so they *can* form in the absence of life, and their bizarre geometry is some kind of self-organizing behaviour arising out of simple (simple!) physics and chemistry. Nonetheless, in most natural examples, microbes have been implicated in helping, in encouraging, in facilitating the process by which they form—at the very least by providing a ready supply of sulphide to link with the iron.

And as for the iron and sulphur, the raw materials—they will probably have had *quite* different journeys before they came together in this shiny new mineral. The sulphur, like the carbon, may have travelled far and wide before finally coming to rest. Take one atom of it as an example: erupted, say, from a volcano as sulphur dioxide gas half a world away and many millions of years before the sediment of the pebble was wrested from its original ancient bedrock. It would have drifted through the atmosphere, then eventually dissolved as sulphate into the ocean. Once in the ocean, that sulphate may have drifted in the currents for millennia, circumnavigating the globe many times, before being dismantled by an energy-hungry microbe and converted into that fool's gold.

The journey of its partner the iron, however, was probably considerably more direct. It had arrived with the sedimentary particles, as part of a mineral grain. On the sea floor, a small part of it could dissolve in those stagnant bottom waters, and so could travel through the water in solution—but probably not far. For if dissolved iron meets oxygenated water, up nearer the surface waters, it immediately oxidizes, becomes a particle of rust, and settles back towards the sea floor. So the journey of the iron was probably short, perhaps of only metres (or millimetres, perhaps) rather than kilometres, as it travelled from its source in the mud towards its rendezvous with the sulphur.

There are hundreds—perhaps thousands—of framboids in the pebble. But the pyrite formed in other places, too—and in a way that helped preserve the remains of organisms that are much, much larger than microbes. Just beneath that Silurian sea floor, in the microscopic world of that soft black mud of the pebble-to-be, perhaps only a few centimetres down, there were spaces, hollows, chambers, caverns. And, in a more outrageously cartoon-like fashion than in any Aladdin's Cave, many of those soon became lined thickly—indeed, often completely filled—with the golden crystals of pyrite. Large crystals, small crystals, clusters of framboids: they could grow extravagantly, creating microscopic golden stalagmites and stalactites in these underground caverns.

These caverns are the insides of fossils: particularly of the baroque interiors of the graptolites, and sometimes of the chitinozoa and acritarchs too. The mechanism of pyrite production is exactly the same—that of breaking down sulphate to produce sulphide, and bringing that into contact with chemically reduced iron in solution, the energy-hungry microbes, teeming in these muds, being the midwives and catalysts of this process.

It is astounding just how effective this chemical factory is: the chambers inside the graptolites can be sometimes 2 or 3 mm across, and they can be completely filled with shining mineral (Plate 2E). Just how can so much iron and sulphur be transported, though glutinous mud (not usually thought of as the most permeable of substances), into and through the cavernous interiors of these fossils?

It is a conundrum. Perhaps, to try to understand the processes at work, one might try to imagine oneself reduced down to molecular size, as in the old sci-fi epic *Fantastic Voyage*, and feeling on one's skin the forces that drive these chemical ions. At that scale, smaller even than that of the microbes, the water is as sticky and viscous as treacle. Will it be quite still? Probably not. Over timescales of months and years, there will be a gentle updraught, a gigantic—at that scale—underground flow, a continuous fountaining as the muds beneath our pebble layer (that total many hundreds of metres thick) are squashed and compressed, simply under their own weight. The water, with its freight of dissolved ions, is forced upwards between the ever more tightly compacted mud particles, eventually finding its way back to merge once more with the seawater above. That is one way to effect a transfer of chemistry within a mud.

At much smaller timescales, there is also that phenomenon known as molecular diffusion. The dissolved molecules, simply by virtue of their own thermal energy, are constantly vibrating, colliding, moving through the fluid like manic dancers moving erratically across a crowded dance floor, or a free-for-all at a dodgem car rally. Colliding with each other, they alter each other's courses, each pursuing a random path through their submicroscopic three-dimensional world.

The collisions are energetic enough to constantly agitate larger particles, such as clay particles or dust grains. Microscopists are familiar with this unmistakable movement, called Brownian motion. It is named after the botanist Robert Brown, who died, after a long and busy life, just before Charles Darwin received news of Alfred Russel Wallace's independent rediscovery of the evolution of animals and plants by natural selection. Indeed, in one of the stranger quirks of science, it was Brown's death that provided the vacant slot in the Linnaean Society's programme that allowed Darwin (finally spurred by Wallace's findings to put an end to his decades-long procrastination) to famously describe his theory (and, graciously, Wallace's too) in public.

Brown was celebrated in life, too, as a Scotsman who could pursue scientific enquiry 'with constancy and a cold mind'. That phrase, one hastens to add, refers to level-headedness rather than coldness of character, for he loved discussion over dinner and port, and once teased Darwin over some new observations of the streaming of cytoplasm in cells ('it's my little secret!' he said in response to Darwin's eager questioning). Brown was not the first person to see the phenomenon associated with his name, but he did describe and interpret it with characteristic thoroughness. He was a superb microscopist, good enough not just to observe and describe the microscopic parts of plants, including pollen, but to look *inside* pollen grains. Within these, he observed even smaller granular particles that seemed to be living a life of their own, jittering and dancing in constant motion.

Was this the activity of life itself, the vital force? No, for Brown saw the same activity in pollen grains that had lain dead in a museum for a century, and then also in inanimate mineral dust too. He didn't know what caused the effect, but he knew and clearly stated that the force lay in the realm of physics, and not of biology. It was later workers (including Alfred Einstein) who revealed it as an effect of the continuous random cannoning of the invisible water molecules around the particles. For a microbe and virus, of course, it's never a quiet life, as they are constantly peppered by this molecular bombardment—indeed, Brownian motion adds a considerable 'random walk' element to their motion that allows them to more effectively chance upon nutrient-rich areas in their environment.

In this nanoworld, Brownian motion is one of the governing forces of both biology and chemistry. Diffusion is highly effective at spreading ions and molecules through a fluid. In the case of the pyrite crystal gardens, growing in the graptolite interiors, there is probably another driving force. Diffusion is a good way to mix molecules evenly through a fluid. When something changes that, to produce areas of high and low concentration, then the molecules and ions, simply by their crazy random walks through the fluid, will eventually balance their concentration levels evenly again. Next to a growing pyrite crystal, iron and sulphide ions are continually being pulled out of solution to snap into place in the molecular crystal lattice. That means that the fluid next to the crystal becomes depleted of iron and sulphide ions, and so more of them migrate in from the fluid beyond, to be snapped up in turn by the growing crystal...and so on. It is a kind of con-veyor belt to bring the raw materials to construct a statue, the internal cast of a fossil.

This is rapid construction. Geologically, it is instantaneous. Often, the fossil is entirely filled with the mineral before its delicate structure is compressed and flattened by the weight of the mud layers above, as, brought in by turbidity currents from the erosional destruction of Avalonia, they accumulate on the sea floor on top of the fossil. Sometimes, though, crystallization takes a little longer, and the sediment has piled higher and heavier. The fossil then begins to distort and deform, before sufficient pyrite forms within it to buttress its delicate walls against the pressure of the growing sediment mass. This, too can be seen in the fossil-cast, which is no longer fully three-dimensional but is now partly flattened—frozen in position at exactly the moment at which sufficient mineral formed to protect the fossil from further pressure. The pressures, as we shall see, will come to be quite enormous. Yet the pyrite will bring the graptolite safely through all of these, for millions of years to come. It is a fossil's best friend.

And a palaeontologist's best friend too, for it renders the fossils supremely visible (and quite beautiful too), the golden sheen contrasting with the grey of the rock, once the brittle carbonized film of the fossil material itself has flaked away. The fidelity of the process can perhaps be seen most remarkably if the pebble is X-rayed, where the graptolites show through as ghostly dark shadows, each living chamber neatly, evenly infilled with the dense pyrite, all through the colony. Indeed, the physical excav-ation of such fossils from the rock is so difficult and so time-consuming, that the instant view of such a fine crowd of fossils comes almost as a shock. A surprise, too, often, for the X-rays pick out fossil species so slender and delicate that they were simply overlooked on the rock surface, even after weeks and months of close and patient examination. But then, surprises are the speciality and stock-in-trade of the Earth.

CROSSING THE LINE

A few millimetres beneath the fine, almost parallel lamina stripes of the pebble, one might glimpse a faint dark shadow, darker than the grey of the mudrock. Looking at the scattered pebbles around, others bear a similar stripe, each just a few millimetres thick—not sharply defined, but fading into the normal mudrock colours. Walk higher up the beach, to where pebbles have been flung by the waves to be exposed to the wind and rain for years—and here those stripes are faded, turning paler, some turned almost completely white.

Look with the magic eye of a scanning electron microscope at a finely polished surface of such a dark—though white-weathering—stripe. The tiny grains that make up the mudrock are seen to be held in a vice-like grip by a bright cement. It is not the kind of lime-based cement that we use to hold our buildings together, but one that is made of the mineral apatite, which is chemically calcium phosphate—the same stuff that our bones are made of. The apatite formed as a kind of rapidly hardening microcrystalline gel a centimetre or two below that sea floor, and it marks a boundary that the pebble mud crossed as it began its slow descent into the subterranean domain—a boundary that was crossed roughly at the same time as the graptolites were beginning to fill with fool's gold.

This boundary is one that occurs now, all around the Earth. It marks the plane—the redox boundary—that separates the oxygenated realm of the surface, with the chemically reducing anoxic realm that lies beneath it. It often lies just beneath the surface. Dig today into a beach sand or mudflat, and a spade's-depth or two below the clean, pale surface sediments, there are usually dark and frankly foul-smelling deposits, suffused with the bad-eggs reek of hydrogen sulphide. At this chemical boundary, minerals concentrate. Just below today's ocean floor, uranium enrichment is often found at this horizon, for this element is relatively soluble when oxidised, but insoluble in reducing conditions, and so precipitates at the redox boundary.

On the Welsh Silurian sea floors, phosphate precipitated at this boundary, derived from the abundant decaying organic matter in the buried sediments. The apatite layers formed most abundantly beneath those sea floors where there was sufficient oxygen to allow the burrowing worms to colonize; that gave the greatest chemical contrast with the stagnant muds just below. But they can also occur just below the laminated layers of an undisturbed 'anoxic' sea floor, presumably because there was still significant contrast between that inhospitable surface and the yet more strongly reducing muds just a few centimetres below that surface.

So, are these apatite-cemented layers relics of the opposition of two chemical realms—a series of redox battlegrounds, successively buried and then re-forming anew a little higher, as each successive turbidity current spread yet another carpet of mud over the sea floor? It probably wasn't that simple. Marine redox layers today are surrounded by a core of mystery, a mystery explained perhaps by the extraordinary abilities of super-communicative microbes.

As conditions above modern sea floors change—say the seawater becomes more or less oxygen-rich—then that change is transmitted to the sediments a few centimetres below the sea floor. That may seem normal, and explicable by simple diffusion of chemistry from seawater to sediment. But the speed at which the buried sediment begins to react to changes in the water above is highly abnormal, because the changes in the sediment, upon being closely observed, are far too fast to be explained by simple diffusion. So something is relaying those changes at lightning speed down into the sediment. And the best candidate for that mysterious 'something' is bacteria acting in concert, forming a chain of electrochemical communication across something like *ten thousand* of their body lengths (to some organisms, a centimetre is an abyss of space).

It's another example of the intricate and sophisticated microbial networks that surround us—and support us—on every side. They are things of which we are almost always blissfully unaware; most of them we haven't discovered yet. Were such quickfire microbial relays integral parts of the Silurian apatite factories? I'd be surprised if they weren't. For microbes, unlike the clumsy latecomer multicellular organisms, have had three billion years to perfect their command and control systems. By the Silurian, they were already ancient and very, very finely honed, capable enough, for sure, of leaving a shadow on a pebble that can endure half a billion years.

THE INVISIBLE METHANE

Let us move on, say, 10,000 years. That is roughly the time that separates us from the end of the last Ice Age, or five times the span that separates us from the Roman Empire. It is but an instant here, at the beginning of the pebble stuff's long journey down into the Earth. Not to its centre—it will not get anywhere near—but it will go far enough for the soft mud to be converted to rock solid enough to line one's roof with. And much, much later, it will come back to the surface again.

But here, it is now some 10 m below the sea floor, which has been covered with a few hundred individual, centimetres-thick mud layers from turbidity currents, and several thousand dustings of fine mud from the slow-moving nepheloid plumes, not to mention the many millions of tiny decaying corpses that have settled from the sunlit waters high above. The microbes are still there in the pebble stratum—albeit they are different microbes, for that second-best source of energy has now gone. The fluid between the flakes of mud and the fine grains of sand is now exhausted of sulphate and so the fool's gold factory has shut down, and another one has started. There are fewer microbes now, and they live and grow and reproduce more slowly, for life here is harder.

The main source of both food and energy is now the buried organic matter. Not all of it. The well-nigh indestructible material of the graptolites, the chitinozoa, and the acritarchs is almost unchanged still, although it is probably beginning to transform, turning from its original transparency, to start acquiring a yellowish tinge, like polythene left too long in the sun. No, the energy source is the other material, the once immensely complex machinery of living cells, carbohydrates and fats and proteins, RNA and DNA too. It is now wreckage, a microscopic junkyard, organic molecules that are decayed, have been digested time and again, each time becoming simpler and smaller, more carbon-rich and poorer in hydrogen, oxygen and nitrogen—yet still capable of forming a snack for generations of highly resourceful bacteria.

At this depth, the bacteria *ferment* what is left, breaking it down further and releasing carbon dioxide and methane, that leak back slowly towards the surface. Even in this, they are picky, selecting the heavier isotopes of carbon to put into the carbon dioxide, and concentrating the lighter isotope in the methane. Not all of these gases reach the sea floor—not immediately at least. Some of the carbon dioxide might encounter alkaline, calcium-rich patches in the buried muds, and react with these to grow hard concrete-like nodules of calcium carbonate, that may be tens of centimetres across—though these are rare in these notoriously calcium-poor Welsh mudrocks.

The methane may well have a more involved history around the pebble—though one that has left no discernable trace—simply a narrative (a genuine ghost, this, so far quite intangible) that we have to infer as being *reasonable*, based upon how this gas behaves in deep sea muds today.

Under parts of the present sea floor, some hundreds of metres below the surface, methane stops being a gas. If the pressure from the weight of overlying mud and water is high enough, and it is not too warm, then the methane solidifies. It becomes a dense waxy substance within the sediment, in which methane molecules are trapped in a cage-like structure of water ice. If you drill some out and bring it to the surface, it un-solidifies at atmospheric pressure, fizzing and cracking and popping—and burning too, if you strike a match near it. It's called a methane clathrate (or methane hydrate). For contemporary humans, it is both a promise (it is a potential fuel) and a threat (if large areas of clathrate are destabilized, in a warming climate for instance, then they can potentially release large amounts of methane—a potent greenhouse gas—into the atmosphere).

Subsea floor layers of methane clathrate today are not fixed. They slowly migrate to stay in the zone where the temperature and pressure are just right to allow it to exist as a solid. As sediment builds up on the surface, the pressure is increased, allowing the clathrate to extend farther and farther upwards in the pile of accumulated muds. But this also pushes the lower part of the clathrate into regions that are warmed by subterranean heat—and so the clathrate turns back into a gas, and migrates upwards into cooler regions where it can then escape, or re-solidify. Thus the whole methane hydrate 'layer' migrates upwards to keep pace with sedimentation, to stay at more or less the same distance below the sea floor.

As our pebble patch of mud was ever more deeply buried, to slowly descend deeper into the Earth, did it spend a few tens or hundreds of thousands of years in a clathrate zone, suffused with a waxy mass of solidified methane? This seems possible, for these Welsh muds were rich in organic material and would have produced methane in large amounts. Can we tell from looking at the rock? This is trickier, for there are few unambiguous clues to the ghost of clathrate past. Perhaps some disruption of the layering? Possibly—if the formation and then disappearance of the clathrate caused expansion and then contraction of the sediment mass. Certainly, disruption of the strata is seen quite often in the Welsh mudrocks, for example where one layer has slid over another while the sediment was still soft—but then such features can also form if the surface sediment slides across a sloping sea floor due to the simple effect of gravity, without any methane being present at all. We have here another little enigma of pebble history, to be solved (hopefully) by further study.

Beneath the clathrate zone, the millennia pass by, one after another. After, say, 100,000 years, we can take stock. The pebble is now buried beneath more than 100 m of mud and sand. It is distinctly warmer, by about 3°C. The mud is more compressed, as a good part of the water has been squeezed out of it, but is by no means yet a hard rock. If one could somehow reach back through time and space to pluck it out of its burial place, it would be plastic and mouldable in your fingers. It is still alive—but the microbes now are much fewer, and live off the ever-more-indigestible food (including their deceased neighbours) much more slowly, dividing now perhaps once a year, rather than once every few hours. Fluid is flowing through it still, very, very slowly upwards.

One can imagine now its shape in the buried strata, taller than the pebble in your hand still, by about 50 per cent, and fatter by about the same amount. The fossil graptolites and chitinozoa and acritarchs have a slightly more yellowish tinge than before, but are still translucent. Some are filled with pyrite, as brittle and shiny and golden still as when it formed. Some are only partly pyrite filled, and here the remaining space inside the fossil is filled with the same fluid as still permeates the compacted mud. And so it will go on, millennium after millennium, as the pebble stuff inches down. With infinitesimal slowness, the mudrock becomes stiffer as more weight is piled on top; and warmer too, by one degree every few tens of thousands of years. The living microbes are slowly thinned out, their glacially slow lives are winking out, one by one (though it will take many thousands of years yet for all of them to finally expire).

Other changes are happening now, so slowly that you barely see the changes from one millennium to the next. The mineral matter of the pebble is beginning to transform. For in amongst the sedimentary debris are swept-in minerals that are uncomfortable in this warm, water-soaked environment. The feldspars, for instance, originally magma-forged aluminosilicates of potassium and sodium and calcium, are slowly breaking down, are decaying. The warm waters, slightly acidic from the remaining organic matter, are slowly dismantling the molecular structures, leaving the aluminosilicate frameworks as new clay minerals, grown within the strata deep under the sea floor. The ions that are released, of potassium and sodium and calcium, are carried away in solution by the slowly coursing fluids.

Some do not travel far. The potassium, in particular, can react with other minerals, such as the magnesium-rich smectite clays, to begin to convert these into the clay mineral illite, which is potassic. This is just the beginning of the long process of the underground transformation of clay minerals that will carry on for millions of years, until they, and the rock (including our pebble) are wholly transformed.

At this stage, we can tiptoe away for a few hundred thousand years, and more, for nothing much new will now happen—that we are aware of, at least—until the pebble reaches a certain phase of, one might say, maturity. And in attaining this maturity, surprising things happen. Let us return to greet the end of the pebble's adolescence: the end of mud and the beginning of rock.

The oil window

It is a few million years later—perhaps three, perhaps five. Sediment has been pouring onto a Silurian sea floor that will, much later, be sliced into by a different sea and become the rugged cliff-fringed coastline of central Wales. It has been pouring in so thick and fast that our pebble stuff is now some two kilometres or more down below that sea floor.

This is quite rapid burial, even by geological standards, and one can blame changing geography for that. To produce a lot of sediment, there is need for a lot of erosion, and also for the production of something that can be eroded—that is, uplands and mountains on land. On Earth, such production of topography is supplied by the marvellous machine of plate tectonics. And at that time, the ocean between Avalonia and Scotland, that we call the Iapetus Ocean, had just about closed, and those two landmasses were just beginning to nudge into each other.

Soft collision, it's called, when the pressure from the adjoining continents is just enough for sections of crust to begin to be pushed up and (to compensate) pushed down in different places—but not enough for the wholesale crumpling that goes with the creation of great mountain ranges. Thus, the landmass that was then in, and just south of, what is today South Wales was driven upwards, while the floor of the sea that then covered Wales was forced downwards. The resultant flood of sediment was Nature's means of trying to restore equilibrium.

Here, the particular pattern of squashing of the pebble stuff is linked with those enormous, mysterious movements of continents hundreds and thousands of kilometres away. And mysterious they certainly were, for on the heels of the soft collision should have followed the hard collision and mountain-building. But it didn't. The mountain-building did take place—but only eventually, and not until many millions of years later. The Welsh mountains are quite a bit younger than they should be—and so that story will have to wait. Another story developed in splendid isolation from such tectonic violence. We can showcase it now.

At a couple of kilometres depth, the temperature of the pebble-to-be is now nearing 100°C and increasing. There is still some water in the spaces between the tightly packed mud particles, but it is not boiling, prevented from doing so by the high pressure generated by that gigantic blanket of sediment.

Is there still life? The subterranean microbes are probably here at their limits of existence. Not that 100°C is a barrier for microbes at the surface, for there are some among the 'extremophile' bacteria that can survive temperatures of at least 113°C near the surface, around submarine volcanic vents. But here at depth in the rock, the combination of heat and lack of food is probably becoming too much even for those resilient organisms. Biology is being left behind, and physics and chemistry are taking over absolute rule of this realm.

And strange chemistry it can be, too. Examine the pebble now, very, *very* closely, with a magnifying glass. Focus on the dark layers, those with the fine striping that record the slow settling of dead plankton on that sea floor. Here and there will be a dimple, barely visible, at best a millimetre across. It marks a tiny patch in the rock that is slightly more resistant than its surroundings. It seems as if it might be anything—a cluster of silt grains, or a cluster of pyrite framboids—something quite normal, anyway.

It is not normal. Take the pebble into the chamber of a scanning electron microscope (it should just about fit) and focus the electron beam on it, set to analyse the chemistry of the tiny dimple and not its shape. Here there is phosphorus, which is a little unexpected, though what occurs with that phosphorus provides the real surprise. There is lanthanum, and cerium, and neodymium, and samarium. Europium too, and gadolinium, and (here the machine is operating at the limits of its analytical capabilities) ytterbium, lutetium and . . . It is quite a crowd of elements. They do not often turn up in everyday conversation (although we have made acquaintance with a couple of them, samarium and neodymium, earlier), except perhaps among chemists of obscure specialization.

These are all rare earth elements, and they have combined with the phosphorus to make the mineral monazite (Plate 3A). There is definitely something strange going on here. The rare earth elements are a singular crowd, and a little inaptly named. They are neither terribly rare (as platinum and osmium, say, genuinely are), nor are they obviously earthy. But they are certainly chemical elements, and they do occupy a row of Mendeleev's fine discovery, the periodic table, being as tightly knit as the closest of human families. Their properties are remarkably similar to each other, which is why they are often found together, when they are found at all.

Monazite is not a rare mineral, as such. It is most typical as a minor mineral of granite, as one that has crystallized out from a cooling silica-rich magma. But unlike, say, calcite or pyrite, monazite was not known as something that forms as crystals in near-surface conditions—and two kilometres down in a pile of compressed mud really does count as near-surface, in the grand scheme of things. That is because the rare earth elements are normally about as near to insoluble in water as one can find. Therefore, one would not expect them to be dissolved out of sedimentary particles, and then to precipitate out as crystals elsewhere in the rock.

So are these monazites simply crystals eroded out of a granite, as are the zircons? Well, no, because they are big as crystals go—up to a millimetre across, and they are also heavy. One would expect to find such large, dense particles in the coarse sandy layers of the pebble, brought in by the strongest currents—and not in the fine, plankton-rich ones. Look more closely with the electron beam, and there appear many impurities in the monazite crystal: particles of clay and silt. These monazites, therefore, are not eroded fragments, but *did* crystallize inside the mud, and engulfed sedimentary grains as the new mineral slowly grew around them.

This is quite a chemical conundrum. Our pebble contains clear evidence that elements that are normally highly insoluble were indeed somehow dissolved within the mud strata, and then crystallized out of solution to form a mineral that is normally associated with magmas. For geologists this is doubly unfortunate. The rare earth elements have long been regarded as so insoluble at the Earth's surface that when they are, say, eroded from a mountain and carried thousands of kilometres within sediment into some deep sea mud, they won't be sorted or segregated, but will retain a chemical memory of that original mountain. They are thus used widely as tracer elements, to work out where the grains that compose rock strata have come from, and to determine the ultimate origin of rock from the Earth's mantle—as we have seen in Chapter 2. To have them moving around within mud strata with such abandon means that the rare earth elements can be reshuffled, to create different and therefore highly misleading chemical patterns.

Those different patterns, in fact, are immediately observed within a single monazite crystal. That electron beam, moving across from the centre of the crystal to its edge, half a milli-metre away, shows that the proportion of the different elements changed dramatically as the crystal grew. In the central, first-formed part of the crystal, the heavier rare earth elements (those with more protons and neutrons in their nuclei), such as gadolineum and dysprosium, are common. Moving outwards, the amounts of these heavy elements fall away, and the edge of the crystal is mostly made up of the lighter rare earth elements such as cerium and lanthanum. As the crystals grew, therefore, the composition of the fluid from which they crystallized radically changed. Within the crystal is captured an archive of the changing conditions at the heart of an enormous mass of mud deep beneath an ancient sea floor.

But how does this wholesale chemical migration and shuf-fling take place? More clues are needed. One can, say, go back to the X-ray of the pebble. And there, among the ghostly images of the graptolites, are the monazite crystals: tiny pale pinpricks, more opaque to the X-rays than the surrounding rock. These, like the graptolites, occur in a swarm in the dark, plankton-rich layers. By contrast, very few monazite crystals occur in the paler bands of mineral mud, those that were laid down by the onrushing turbidity currents.

So there's one clue—the monazites somehow have an affinity with the organic matter of the plankton-rich layers. But where did the rare earth elements come from, to form all those crystals? One might here sacrifice the pebble (once more!), carefully cutting out the dark, carbon-rich layers and the pale, carbon-poor ones, and making separate chemical analyses of each, to see how rich each is in the rare earth elements. The pattern is striking: the dark layers are just stuffed with rare earth elements, at ten times or more the amount seen in an average mud. The pale layers, though, are markedly depleted of them, containing smaller amounts than are present in average mud. Put the two together, and the rock as a whole contains normal, entirely unremarkable amounts of the rare earth elements.

The process, therefore, must be one of wholesale redistribution, transporting these elements from the organic-poor to the organic-rich layers, deep inside the rock. What does the transporting? Well, one factor must be the water, now scaldingly hot, still coursing up through the entire mass of mud, filtering very slowly through the now tightly packed sediment particles, as the mud squeezes itself ever closer to dryness under its own enormous weight. This can dissolve material from lower layers, carry it up in solution, and then the material may be precipitated out once more in higher strata, where conditions are different.

Something like that seems to have affected the rare earth elements. But these are, if you recall, normally highly insoluble. Is there some kind of factor X involved, to help them on their journey?

THE OIL PROVINCE

There was certainly another, very considerable phenomenon taking place then. The evidence is a bit circumstantial as yet, but one suspects that it had a hand in this remarkable mass transport of normally untransportable chemical elements, and perhaps in much else. For the pebble stuff, and all the strata around it, was then going through the oil window.

Oil is the stuff that has most shaped our lives over the past century. It is the most marvellously convenient energy source, easily piped out of the ground, transported, controlled—the fuel of choice for a modern civilization. There are downsides, of course—all that carbon, once it is burnt, has to go *somewhere*. But the siren song of oil has been so sweet that such consider-ations have, so far, taken a back seat (when they have taken any kind of seat at all). And so oil has been the subject of deep interest to many people, not least as regards the manner of its creation, a long time ago, deep underground.

The pebble has produced its share of oil—perhaps a fraction of a teaspoonful. Its production, like that of the monazites, also started at somewhere between one and two kilometres underground. The temperature, then, had risen suf-ficiently to simmer the remains of the plankton in the mud, slowly and gently. The organic molecules, already degraded by the busy microbes from the complex and extravagant molecules of the living organisms, broke down even further. Fragments broke away from them, which were of just the right size to be in liquid form. There's quite a variety here: from the stick-like alkanes to the circular cycloalkanes, with a sprinkling too of the 'aromatic' compounds—those that contain the distinctive hexagonal structure of the benzene ring. Smaller fragments broke off, too, taking the form of natural gas—mainly methane, but also some ethane, propane, and butane.

It is a natural distillation that goes on for a few million years, though it is easier to measure it in depth and temperature rather than in time. Oil and gas are slowly released from the fossilized plankton remains, as the pebble strata slowly descend from about two to about five kilometres below the sea floor, as yet more sediment is loaded on top of that subsiding sea floor and the temperature rises from about 80°C to about 150°C. Beyond that depth, that's pretty well it for the oil—the plankton have released as much as they ever will, though some gas will be given off for another kilometre or two yet; and then even that will finish. The rock has then passed through the oil window, and the gas window too, and that is the end of production. What is left is what is visible in the pebble now—blackened carbon husks, for the most part as graphite and amorphous carbon. If you analyse the pebble now, it contains somewhere between one and two per cent carbon. The original mud likely had ten per cent or more, and a considerable part of this went into forming oil and gas. Nature's providence, indeed.

Where did it go, that oil? It travelled upwards, and not just because of the squeezing of the mud, but because it is less dense than water, and in any mix of the two, it will rise towards the top. If it encounters strata in which the spaces between the grains are larger—in layers of sand or sandstone, say—then its travel is much easier through those roomier subterranean passageways. The oil and gas might, perhaps, travel all the way back to the surface, breaking through somewhere on the sea floor as a hydrocarbon seep. Such seeps today nourish communities of animals and plants for which the return of long-buried carbon is welcome life support. These communities are the first step in putting that oil and gas back into the carbon cycle of the Earth's surface, from which it has been absent for some millions of years.

The question of how the oil travelled, though, is a tricky one. Oil is, by repute and in reality, thicker than water. It forms tiny elongated droplets—the oilmen call them slugs—that somehow have to wrestle and squeeze their way up through what are now infinitesimally small gaps between the tightly packed flakes of mud. It is an escape trick worthy of Houdini, and it is still uncertain quite how this is done. Oil is forming and migrating today—for instance deep under the Gulf of Mexico—but it is a difficult environment to monitor closely enough (and patiently enough!) to detect and record exactly how the oil travels. One idea is that it partly blasts its way out, the pressure from the oil's partner, the gas, helping to temporarily wedge open tiny pathways between the grains to enable the oil slugs to squeeze through to higher and higher levels. It's possible.

Gaze at the pebble and you are gazing, still, at one of the many mysteries of the Earth. For not only did that pebble yield up its own few drops of oil, but it also allowed through it many more such droplets that travelled up from strata below, on their way towards the surface. They may not have passed through completely without trace; part of the carbon coating those sedimentary grains is probably a residue of oil from much deeper down, that was simply passing through. Building a pebble is not—one sees yet again—a simple process.

Alternatively, the oil and gas—including the pebble's tiny contribution—might end up trapped in some subterranean blind alley. It may have entered a thick bed of sandstone from beneath, say, that was entirely covered by a layer of mudrock too thick and too impermeable for it to penetrate further. This is now an oil reservoir. Then it stays underground, until the rocks above it are eroded away and it, too, finally breaks through to the surface, perhaps hundreds of millions of years later. Or perhaps earth movements will break the seal before then, allowing the oil to gradually work its way back upwards again. Then again, the oil might not survive entirely pristine in its underground tomb. If that tomb is carried too close to the surface, within a kilometre or so, then it comes within range of those hungry microbes that can, yet again, extract some nourishment from it. This turns the oil sour and unusable, at least to certain geologically recent bipedal apes.

These apes can get no more use, in any event, out of Welsh oil. In the times of the coal swamps and the dinosaurs, this may well have been an oil province to rival the Middle East. After all, the main oil source rock in Saudi Arabia is a Silurian organic-rich mudrock laid down in an anoxic sea, the exact equivalent of the strata of central Wales (and the Arabian graptolites are, by the way, beautiful). Here, though, the oil reservoirs were broken up and eroded millions of years ago. Perhaps not entirely: oil companies have occasionally drilled speculative boreholes around Wales, hoping—so far in vain—that some last small pockets remain.

But the great escape of the oil may have left other memories in the pebble, memories of the kind that are avidly sought by geologists, those that bear a date-stamp, to tell us when the escape took place. This date-stamp is still a little hopeful: work in progress, one might say. It hinges on the connection between the escape of the oil and the role that this might have played in that other great chemical reshuffling: the migration of the rare earth elements and the crystallization of the monazite in the strata. We were looking, you might recall, for a factor X to help transport those rare earths through the muds. That factor might just have been the oil migration through the rock.

The rare earth elements may be obscure, but they have not been unstudied. Recently there has been a surge of interest in them because they are a key ingredient of certain high-temperature superconductors. 'High temperature' here might be a hundred degrees *below* zero, for superconductors normally operate at much colder temperatures still. If superconductors can be made to work at room temperature (or at something like this), then they could allow the much more efficient use of electricity. And if the rare earths are a key ingredient of these, then the monazites of pebble type might, perhaps, become a sought-after, prospected-for and marketable commodity.

Among the lesser-known byways of research, though, has been investigation of what might make rare earths dissolve more easily in water. One thing that has been found to make them more soluble is the presence of organic matter in the water, with the hydrocarbon molecules grabbing onto, or 'complexing' with, rare-earth atoms. Such mobile organic molecules will have been abundant as the oil was moving, helping the rare earths piggyback those few centimetres from the pale turbidite muds into the dark plankton-rich mud layers—where they stuck, and stayed, and formed crystals that are still visible in the rock (and the pebble) today.

TIME AND TIME AGAIN

The monazite crystals have time built into them. Alas, this is not simple and precise time of the kind that those marvellous zircons contain. This is tricksy time, time that tries to catch you out, time that is subtle and that has (one might almost say) a sense of humour—and a rather offbeat sense of humour, at that. Each monazite crystal contains two clocks. The first of these is the more obvious—and is also downright misleading. One of the rare earths—samarium—has a radioactive isotope that decays at just about the right speed for a deep-time chronometer into another rare earth, an isotope of neodymium. Both are present in monazite, in reasonably large amounts.

Bingo!—one might think—the perfect chronometer for mudrocks. Unfortunately, there's a catch. One can try this out by, very carefully, analysing the amounts of parent samarium and daughter neodymium, and from that—knowing the rate at which one converts into another (the half-life)—work out the time since the crystal formed. The answer arrived at, however, suggests that the crystal formed over a hundred million years *before* the mudrock that it formed in was laid down. This is simply nonsensical, and there must be a reason for it.

There is indeed a reason, and it reveals the subtlety—one wants to say cunning—of the monazite crystallization process. If you look closely at the composition of the neodymium in the monazite—that is, the proportions there are of its different isotopes—throughout the crystal, then one finds that there is apparently more of the samarium-generated isotope in the core of the crystal than there is in its outer parts. This extra neo-dymium was added to the crystal when it formed, from some unknown outside source, and it effectively 'wound forward' the clock. As there seems no way yet to work out by quite how much the clock has been wound forward, the clock is therefore useless. It's fascinating and intricate stuff, but it does demon-strate the pitfalls of getting to grips with—*attempting* to get to grips with—the ghost of geological time.

Fortunately, the monazite has another chronometer up its sleeve. Monazite (like zircon) can, when it grows, take in a little uranium. This, by its well-known radioactive decay to lead, can tell us when the crystal formed. It's a close-run thing, though, for our type of pebble-monazite contains tiny amounts of uranium: only just enough to enable measurement of how much of it—and of the lead that formed from its decay—is present. These age estimates, therefore, do not rival the sub-million-year pre-cision of the zircons. Nevertheless, they clearly show that the pebble type of monazite finished crystallizing some 415 to 420 million years ago, which is several million years after the sedi-mentation of the mud that surrounds it. And, if the monazite crystallization was catalysed by the migration of oil, then this clock provides a date, too, for when the oil was generated.

The pebble is a microcosm of a now-vanished underground world of almost infinite four-dimensional complexity. And this particular phase of its evolution was marked by more than changes in what were, after all, minor components: the few per cent of carbon compounds, and the minuscule fraction of it that is made up by the rare earth elements. The entire fabric of the mud was changing too. It was becoming a rock.

In the pressure and oven-heat of kilometres-deep burial, and the catalytic effect of the underground fluids, the grains themselves were altering. The feldspar particles, by now, have disappeared, broken down into clays. The quartz grains are being dissolved at their margins, particularly at those high-pressure points where grain rests on grain (and carries on it the weight of miles of strata above). There, silica goes into solution—only to precipitate in adjacent patches of lower pressure, forming a cement around the grains and binding them together.

The delicate, complex clay particles, too, were transforming. They were growing. Or, more precisely, some were growing at the expense of others. There was a continual rearrangement of the molecular lattices; some were being broken down, while others were added to. The particles, originally numerous, tiny, ragged, untidy, were now transforming into a meshwork of larger, thicker, better organized crystals. It is the gradual beginning of a process that will evolve into the fully fledged metamorphism associated with mountain building, a dynamic process where the crust itself is shortened—but here the strata were simply stewing a couple of miles below a subsiding sea floor, as yet more sediment poured into the sea off the coast of the vanished continent of Avalonia.

One of the particular transformations here among the clays has a far-reaching effect. Or rather, this transformation is the prime suspect for a far-reaching, though entirely invisible, transformation of the rock, and the motive force for yet another chronometer of Earth history—of which the pebble can form a part, but not the working whole.

The transformation here involves the clay mineral smectite. This particular clay mineral is common—it is a standard weathering product of volcanic rocks, for instance. It also contains water molecules, more or less loosely bound into its structure, and these can be absorbed or released as the environment changes. It is not a clay mineral that you want to have too much of in the soil beneath your house foundations, for instance, for in summer the ground will dry and shrink, and in winter it will become wet and expand—and the house will go up and down as this happens.

At depths of around two kilometres, the smectite structure begins to collapse. The smectite typically converts into another clay mineral, illite, by picking up more potassium from the debris left over from the decayed feldspars. The water from the smectite structure is released, and adds to the fluids coursing through the rocks. It is an extra fluid pulse of some significance, for it may explain some very strange patterns that have emerged from the Welsh mudrocks.

One of the more subtle rock-chronometers of mudrocks involves the element rubidium. One of its isotopes is radio-active, and it decays into one of the isotopes of strontium. This chronometer requires the analysis of several samples, usually taken from a little distance apart, so a single pebble won't do, alas: one needs several pebbles—and ones, moreover, originally derived from close together in the rock, which is a little tricky to ensure. If the proportions of the rubidium and the strontium line up nicely as an isochron on the ensuing graph, then that can give the time since the isotopes involved were last mixed together evenly. That mixing event is the reset time in this particular atomic clock.

The rubidium–strontium clock had been used, in Wales, to try to precisely date a particular event, one that our pebble-form is (in this narrative) still to undergo: the tectonic event that created the Welsh mountains. This was thought to be the only geological event important enough to thoroughly reshuffle the isotopes in a volume of rock as large as, say, a house.

The rocks were accordingly analyzed. A major reshuffling of the rubidium and strontium isotopes was indeed recognized in the rocks. Its timing seemed wrong, though. The reshuffling event that was revealed by the rubidium and strontium isotopes was almost 20 million years older than other estimates of the mountain-building event had suggested. It took place, indeed, at much the same time as the monazite crystallization event. Had the mountains formed much earlier in this part of Wales? This is unlikely (as we shall see). So did something else reset the rubidium–strontium clock in the rocks, so thoroughly that even the subsequent mountain-building had no effect? That was held to be a more reasonable interpretation, and the finger was pointed at the pulse of fluid released by the pressure-driven dehydration of smectite in the rock as it transformed into illite.

It is the most subtle of effects in one sense—invisible, intangible—but nevertheless it caused thorough mixing of one facet of chemistry, on a scale that once more underscores the dynamism of this strange world of the underground. It is a world that is very close to us. All we have to do to reach it is to walk a few kilometres vertically downwards. If the road were clear (and we had the right kind of shoes, the circus-type ones that can cling to a vertical wall), it would take us less than an hour. But then, to really see what was going on, we would need to develop feelers to detect the subtle movements of isotopes, of fluids and of how they change; we would need pressure and temperature sensors; and we would need time, to observe and to record, and then to think through those observations and recordings, for millennium after millennium. It is not a task for a human mind. Luckily, human impatience is better served by the fossilized relics of these long histories preserved in strata—and, remarkably, that are encapsulated in our single pebble.

At this stage, the pebble-shape in the strata is now a rock. If one could reach back through that time and space, and pluck it out, it would be approaching its present shape. It would be about the same height as it is now, but it would be fatter, still, than now. At this point, it has mightily been squeezed from one direction—from above—but not yet from another. Hit it with a hammer and it would break with a clink, cleanly along the sedimentary layers, sometimes exposing those beautiful pyritized graptolites. (Oh! If it could but do the same today, a palaeontologist's life would be so much easier.)

We need to leave the pebble now for some time, there in those dark depths. For now nothing much will happen to it for almost twenty million years, except that the clay minerals will slowly grow larger and more tightly bound together, and the rock will get harder. It is trapped in those depths. It awaits one more transformation, as a continent approaches from the south. Africa is set to leave its mark on our pebble.

Making mountains

THE SLATE MAKER

It has been a quiet 20 million years for the pebble: an interlude, at somewhere around 3–4 kilometres under the sea floor. The rock has still been crystallizing, but only very slowly. The water has by now mostly been squeezed out, so little fluid has flowed through that rock. At this depth it is hot, well above 100°C. The pebble-form is sterile, lifeless.

The time is now a little under 400 million years ago. We are in the Devonian Period. Above, at the Earth's surface, changes have been taking place, but as far as they affected the pebble they could be on another planet. In the sea, the graptolites have been going through an evolutionary rollercoaster, with explosions of diversity separated by bad times, when they only just survive. Soon, one of those bad times will be terminal, and they will disappear from the open seas, never to return. By contrast, the fish are beginning to thrive both in the sea and in rivers and lakes. The land is greening, almost explosively, as plants evolve furiously.

None of this affects the future pebble. But something soon will. The sea above has been gradually shallowing, filled in with sediment from the encroaching land. Eventually, it changed, some few million years ago, into a vast coastal plain, traversed by rivers. We are about at the time, now, when that lowland is about to rear up to form a range of mountains that—although much reduced from their early glory—can still be climbed today.

What took them so long? For the Iapetus Ocean to the north, which, 50 million years ago, was more than 1000 kilometres across, had effectively disappeared 20 million years ago, the ocean plate sliding beneath the northern continent of Scotland and north America. But on Avalonia, the effect was as if these continents had just slid neatly into place, with only minor distortion of the Avalonian crust (and, in truth, these landmasses did approach each other partly from the side, rather than head-on). Did the mountain-building force still come from the north, perhaps as some mysteriously delayed intensification of the vice-like grip that held these landmasses together?

Or did the force, rather, come from the south? The latter may be more likely. For Avalonia's separation from the Africa-dominated landmass of Gondwana, which had lasted some 200 million years, and taken them thousands of kilometres apart, was about to come to an end. For that wide southern ocean between them, the Rheic Ocean, was disappearing too. And the timing of Avalonia's docking against Gondwana—or rather with its then promontory of Iberia (now Spain) and Armorica (France, largely) seems, as evidence accumulates, to coincide with the timing of the Welsh mountain-building.

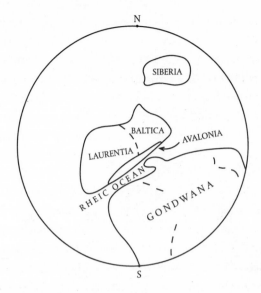

FIGURE 8 The position of Avalonia in the Devonian.

The fabric of the pebble stuff, now, is about to undergo the last of its great transformative changes. It has, for a long time, been under pressure—but that pressure has been simply due to the weight of a great mass of strata above it. That pressure has flattened it and squeezed it almost dry; but, confined as it is within its own stratal level, that pressure has been felt all around it. It is akin, say, to the crushing pressure that an incautious deep-sea diver feels on descending towards abyssal depths. But now the pressure regime is set to change.

As the crustal masses of Avalonia and Africa press against each other, driven by the slowly moving mantle currents deep inside the Earth, the crust of these continents is crumpled and thickened, particularly around the zone of impact. The pebble-to-be feels this as an insistent, directed pressure from the side, as if it were in the jaws of an enormous vice. One jaw is in the north-west, and another in the south-east—that is, mind, as they are measured relative to where Wales is on the Earth today.

The pebble is now moving, following a tortuous path both from side to side and up and down, as the strata that it lies in are thrown into huge folds (Plate 3D). They cannot be seen in the pebble—one has to look up and glance at the cliffs, where individual folds can be as big as the cliff itself. How does one fold such a hard and brittle thing as a rock? The answer is: with a great deal of force (obviously), and also that that force has to be applied *slowly*. If the pressure is applied too quickly, then the rock simply breaks, with the destructive snap that is an earthquake. Such fractures are indeed visible, here and there in the cliff, where the strata have been violently dislocated. But the hot, pressurized rocks underground can also slowly creep and flow, under a steadily applied pressure. It is a little like the case with glacier ice, which can flow downhill or break into crevasses for exactly the same reason—or, more prosaically, with that child's toy, Silly Putty, which can be made to stretch or to snap, depending on whether one smoothly pulls on it or gives it a sharp tug.

It is also getting hotter. This is not so much because the folding is pushing it downwards (although that may happen) but because the whole mass of strata—the size of Wales, indeed—is being thickened as the crust is shortened. This forms a thicker blanket to trap the Earth's heat. The rising land (that was the sea floor on which the pebble-sediment settled) is now narrower from north-west to south-east by 10 per cent or so. By the time it finishes moving, this section of land may be as much as a third narrower. Within this thickened pile of strata, the heat is building up, climbing slowly towards 250°C.

The pebble-form is now, finally, being squeezed into shape—its present shape. Like most of the other pebbles on the beach, it is essentially disc-shaped, reflecting the way the rock split into slabs, the edges of which have then been rounded off by the waves. But the slabs are not strata-slabs, for the stripes of the stratification run askew to the flat surface, rather than parallel to it. The slabs reflect the splitting-surfaces, and these have been fashioned by those growing mountains, as the mudstone is transformed into a slate. One could—if one wished—take a hammer and chisel and split the pebble further. The craftsmen who used to make roof-tiles by hand in the old Welsh slate mines could get them down to a thickness of less than a quarter of an inch. One couldn't do that, quite, with the pebble—the sandy layers interfere with the splitting—and, of course, the skill of the old craftsmen was hard-won and is not easily duplicated. But the individual flakes (that would shower from the clumsily applied chisel-end) may be a fraction of a millimetre thick.

So, in the roots of the new mountain belt, the microscopic fabric of the pebble was transformed too, as the folds grew. The insistent lateral pressure was transmitted through the entire rock mass, acting on the countless flake-like micas. These, up until then, had mostly been lying flat as they fell, parallel to the sea floor. The micas began to change their position, to align themselves at right angles to the forces. It is like pushing with your hand against the edge of a sheet of paper. It will bend until the flat of the paper is against the flat of your hand.

The reorientation takes place along planes that are at right angles to the insistent pressure, with—at least initially—planes of still untransformed micas between these. This planar structure—along which the rock now splits easily—is termed a tectonic cleavage, or just a slaty cleavage (Plate 3E). As the heat and pressure persists, more and more of the micas are reoriented, until eventually they all are. In theory, at that stage of thoroughgoing transformation, you could carry on splitting the rock along the cleavage planes almost to infinity—although one would need all the skill of the old Welsh slate miners, and more, to do that.

The readjustment is complex. The micas can rotate mechanically, or they can regrow, dissolving and recrystallizing to lie flat against the pressure. Or, they can bend and then fold—and once one is folded, the one above tends to fold also, and then the one above that, and then the next, until a tiny vertical stack of folded micas is formed. Geologists call this a crenulation, and it mimics on a microscopic scale the huge folds that can be seen in the cliff.

For all this to happen, water is still needed. Amazingly, after all the squeezing and heating of the past 20 million years, there is still some fluid left in this rock mass—only a few per cent, but enough to allow the multiple episodes of dissolution and precipitation at that molecular scale to effect the rock transformation. Also—and a little more mysteriously—there seems to be enough fluid to spirit away some of the rock up towards the sky.

IMPORT AND EXPORT

Mudrock becomes smaller once it has been transformed into a slate. Some of its mass is lost, which some studies have suggested can be as much as a fifth of the rock material. Part of this is silica, taken away in solution. How does it travel? Again, here we are in the realm of a mystery deeper even than that of the expulsion of oil from a mudstone. The cleavage planes might be possible transport pathways, to allow the upwards export of both fluid and dissolved mineral. If they are pathways, though, they are ones of almost non-existent thickness, and moreover ones that are being pressed together by all the forces of mountain-building. It is another of the mysteries of the pebble. Gazing at its smooth form, the nature of those pathways remains a puzzle, another of the many enigmas still left to decipher.

However this material travels—where does it go? Some of it goes a little way upwards in the rock, to fill mineral veins and fractures, and *that* is part of the next section of this narrative. But some probably travels much farther. In today's mountain ranges, there are springs bringing forth mineralized water that has been derived from very deep down in the Earth's crust. Similar springs certainly issued from those growing Welsh mountains. They would have carried the kind of elements that you can read on the label of a bottle of mineral water: silicon, magnesium, calcium, sodium and more—and perhaps (as was present in one bottle I once incautiously bought) some more exotic elements such as mercury. All this was being squeezed out from the strata beneath, and was all helping to fertilize the new landscape as, for the first time in the Earth's four billion year history, it begins to develop a land-based vegetation. The pebble, too, probably contributed its microscopic ha'pennyworth to the creation of a green and pleasant landscape.

This is a strange environment, within the high-pressure jaws beneath the roots of mountains. It is harder to visit and sample than is (say) the surface of Mars, but the processes within it are crucial in fashioning our landscape and the rocks that make them up. The Welsh mountains, we must remember, show only the outer circle of this inner-Earth realm. Deeper levels may be seen in the rocks exhumed within the larger mountain belts, of the Alps and the Andes and the Himalayas. There, the temperatures rose to 400, 500 and even 600°C, while pressures doubled and quadrupled. The tortured rocks were transformed beyond recognition—some up to the point of melting.

The pebble may not represent deep-Earth processes on such an intense and Alpine scale, but it does hold within its confines myriad clues to just how the Earth's pressure was transmitted through it, and transformed its complex mineral framework. Look, for instance, with a hand lens at its surface. There are dozens of pale specks, about the size of beach sand grains, scattered on the dark mudrock surface. Investigated more closely, under a microscope, these turn out to be large mica crystals. The interior of the pebble contains many thousands of them. These are not, though, like the thin, shiny mica flakes that are frequently washed along river and sea floors, and that give the sediments of these their sparkling appearance. The oddities in the pebble are shaped like barrels, higher than they are wide, standing up on the stratal surfaces (Plate 4A).

These barrel-shaped micas are another product of the mountain-building. Originally they were the normal flat washed-in flakes of mica, lying along the lamination surface. Then, as the rock was squeezed from the side, the sheet-like layers of the molecular lattice were forced apart, allowing chemically charged fluid into the interior of the flake. From this fluid there crystallized more mica within the original mica, making it grow thicker. Then this happened again, and again, and again, until the mica particle grew to ten or more times its original thickness. It is like converting a slim paperback into the heftiest of tomes, by gradually inserting many thousands more pages between the original pages. Where did the material come from, to enlarge these mica 'books'? It must have travelled in from somewhere else in the rock mass, as a kind of chemical import-export, though whether this took place over a range of milli-metres or metres away is unclear.

However, in regions of the pebble that were protected from the crushing sideways pressures, the micas retained their original shapes. The micas, that had been engulfed some 20 million years earlier by the growing monazite crystals, found just such a haven. The monazites acted like steel ball-bearings that are embedded within plasticine that is being squeezed hard by some energetic child. They were not deformed as the mudrock around them was converted to a slate. The micas within the embrace of that rigid monazite remained as thin and flat as when they had settled, a few million years earlier still, upon that Silurian sea floor.

THE SHADOW OF TIME

Other objects inside the pebble also resisted the mountain-building pressures. The pyrite-filled fossils, those golden stick-like graptolites, could not be flattened or bent. They could be fractured, though, and many graptolites in such rocks are broken at their thinnest and weakest parts. The tectonic pres-sures can convert such fossils into isolated segments, which are either separated by small gaps or partially stacked against each other, depending on whether the mountain-building forces have stretched or squeezed them. It provides yet another jigsaw puzzle for the poor palaeontologist, who has to try to put the pieces back together to work out the shape of the original animal. One consolation is that, within each segment, the rigid pyrite has faithfully protected the detailed shape of each chamber from the wholesale deformation that has affected the rest of the rock.

There is another effect of the tectonic squeezing, though, that adds another layer of complexity—both to this narrative and to the fossil. As a bonus, it weaves yet another chronometer into the pebble, that will allow us to say quite when the Welsh mountains—and the pebble—were raised up high. It is a chronometer with a weirdly distorted action (and one for which a little luck was needed at its unearthing). But then the distortion itself seems to be telling us something (we are not yet sure quite what) about how these mountains were raised.

We might start here by recalling the painstaking unearthing of those graptolites from their enclosing rock matrix, with the help of that steel needle mounted in a pin-vice. Each graptolite is coated in a whitish friable material, which may be up to a millimetre thick (Plate 4B). That coating is made of mica, and it is normally sacrificed as the palaeontologist tries to get at the fossil underneath. The outer surface of this mica coating also faithfully preserves the shape of the fossil, but it requires more skill, more patience and a lot of luck to be able to leave this delicate mineral sheath intact as one slowly exposes the fossil (life—one mutters, as one rubs one's painful wrist and shoulder muscles while hunched over the microscope—is sometimes just too short). To a palaeontologist, the coating is a nuisance. But to a geochronologist, aiming to conjure estimates of numbers of millions of years out of the rock, it is gold dust.

There is a simple explanation for the coating. (The simple explanation, by the way, is wrong, but it will do for a start.) When the rigid, pyrite-filled graptolite is acted on by the mountain-building forces, it reacts differently from the mudrock that encloses it. We are back to the scenario of the steel ball-bearing in the energetically squeezed plasticine (although in reality the metal-filled graptolite is more like a stout steel nail). If you try this, and look carefully, you will see that above and below where the plasticine is being pressed hard against the ball-bearing, it actually parts company from that object, leaving a small gap between steel and plasticine. This is what tectonic geologists call a pressure shadow, a protected zone in the lee of the force that is being exerted. This kind of thing happens, too, deep in actively deforming mountain roots, where spaces open in the pressure shadows around rigid objects held within mudrock or slate that is deforming more plastically around them.

These spaces are instantly filled by hot, mineral-rich fluid. And, in the spaces that formed around the graptolites, tiny crystals of new mica precipitated from the fluid. As the pressure continued to be applied, the space grew larger, and became fluid- and then mineral-filled in turn. One can track the growth of the mica crystals by their shape—they typically form crystalline fibres that grew at their ends. The pattern of the fibres mimics the way in which the new mica filled the space: sometimes the fibres can be seen, say, to bend as the graptolite slowly rotated within its micro-cavern, another effect of the tectonic pressures around them.

It is a neat story, this, and it is a shame that it is not true. Or, more precisely, it is not sufficient to explain all the facts, which is a very common situation in geology, and indeed in science as a whole. Ugly Facts keep disturbing the symmetry of a neat explanation—but then the Ugly Facts, once understood and accommodated, can widen and deepen our understanding of how the Earth works, even in its smallest and most obscure corners.

For such mica coatings are present not just on those graptolites that have a rigid pyrite infill. On rocks that have suffered these kinds of pressures, they are present on pretty well all graptolites, even those where all that has survived is a wispy carbon film: a film that could not have acted as a rigid bulwark capable of resisting immense pressures, by any stretch of the imagination. And, some other rigid objects in such rocks, say fragments of shell or crystals of pyrite, do have a mineral coating around them—*but* this coating is not of mica but is of other minerals, most typically quartz. So what is, or was, going on?

There is an extra factor. That seems to be the organic matter of the graptolite, which seems to have had a catalytic effect within this hot, deforming rock mass, causing the crystallization of mica rather than of other minerals. It reflects a peculiar chemical and mineralogical selectivity during mountainbuilding, and underscores the sheer subtlety of the processes at work in this hidden realm. It is a quirk, perhaps, but this evidence of active connection between the organic and inorganic chemical kingdoms may have much wider significance yet.

It is significant for palaeontologists, certainly. Some celebrated fossils, notably the ancient soft-bodied animals of the Burgess Shale of British Columbia in Canada, perhaps the most celebrated window into early life, have such shiny mica coatings. These were originally interpreted as the remains of clays that grew on the fossils just after their death (and so preserved them). The evidence of the graptolites suggests, rather, that the micas grew not just after death but millions of years later, when the carbonized husks of the Canadian fossils catalysed their crystallization deep within the Earth. Hence the micas do not solve the question of why these soft-bodied animals were preserved (though the sheen of this mineral does make them clearly and beautifully visible). To understand why these delicate creatures were preserved at all, one needs to look for other reasons—for example that they may have been buried very deeply and suddenly in submarine avalanches. The pebble can contain insights, therefore, that can resonate around half a world and across millions of years.

Nevertheless, even as such narratives become ever more byzantine and (one hopes and trusts) ever more true, the phenomena they describe may retain a simple utility. And the simple question here is: when did the Welsh mountains form? This has been a difficult question to answer directly. In mountain belts that show more of their high-temperature cores, the minerals have grown larger, and those that contain useful radioactive elements such as uranium can be separated out and analysed, to hopefully yield the date of their crystallization—and therefore of the peak phase of the mountain-building.

No such luck with a Welsh slate. The new micas that define the slaty fabric do contain a potential chronometer: the potassium in these crystals has a radioactive isotope that decays to the noble gas argon. But the crystals are typically so tiny, and so mixed with other, older mineral grains that were washed in as the mud was accumulating, that no sensible way was found of getting a pure enough sample to exploit this particular atomic clock.

However, the mica coatings around the graptolites do seem to represent such a pure source of mineral, that grew in a cavity that could be up to a millimetre in size (*huge* by the standards of such delicate analysis) as the rock was squeezed and transformed by the immense tectonic pressures. The first time this method was tried, it scored a bull's-eye: the mineral had formed 396.1 million years ago, plus or minus 1.4 million years. That was unheard-of precision for a slate—almost rivalling that of the marvellous zircons, and confirming the huge gulf of time that separates the mountain-building event from the deep burial of the strata—and of the pebble (which took place, as the monazite and rubidium–strontium clocks showed, at some 415 to 420 million years ago).

A magic bullet for dating mountain-building? Not so fast. Attempts to repeat and refine the method didn't give such marvellous precision—in fact, some of the results seemed non-sensical. It took some detailed, painstaking work to show why, involving the firing of a laser beam along a set of microscopic, closely spaced points across the whole millimetre thickness of a mica space-fill. This showed that some parts of the mica consistently gave dates of around 395 million years, but between these were rhythmically spaced zones that gave much older dates—nonsensically older dates that stretched back to hundreds of millions of years before the rock itself was formed. They have too much argon in them. That first attempt had, with no little serendipity, sampled pure sections of the consistent material.

It's another case of an atomic clock that has been wound forward—but here this has only happened to parts of it, while other parts seem to work perfectly well. (This is still a clock that can be used, albeit one that has to be handled with great care.) The partial 'rewinding' is the periodic addition of 'excess argon' to the growing crystal, and the rhythmic way that this seems to have been done suggests (perhaps!) some kind of rhythmic action associated with the mountain building, maybe as a pulsing of fluid through the deforming rock mass.

The heartbeat of a mountain belt? Well, it is yet another enigma to add to the catalogue of mysteries still packed tightly into our pebble. Work in progress, once more. And there is yet more to do. Occasionally, in with the mica fibres that surround these fossils, there is a fibrous crystal of very different chemistry—albeit one that has become familiar to us in an earlier context. Monazite has, very rarely, crystallized in this setting, which means that, somehow, enough fluid must have occasionally flowed past these fossils to transport and crystallize those almost-insoluble rare earth elements one more time, as the almost-dry rocks surrounding them were deformed. This potentially represents yet one more atomic clock to use to date the timing of mountain-building. But, monazites of this 'tectonic' generation are tiny, and they contain minuscule amounts of uranium. It would be a daunting task to read this particular chronometer: a challenge for the next generation of atom-counting machines (and scientists), perhaps.

We are now at the furthest point of our journey—and the most extreme, as regards the conditions that our pebble has endured. At this point, it is virtually in the form it is in *now*, as it lies in the palm of our hand—although, of course, it is still surrounded by rock in all directions, reaching upwards for several kilometres and downwards for several thousand kilometres.

It will be surrounded by rock, still, for almost 400 million years more. In that time, it will slowly ascend, as the mountains above it are worn away by the wind and weather, and it will cool. Many events will take place on Earth above it, both near and far, some slow and gentle, some violent and catastrophic. Few will have any effect on our pebble, insulated as it is from the hurly-burly above.

Before its long sleep begins, though, there are still some alterations to be made to its fabric. As it begins its travels upwards, and as the crustal vice that held Wales in its grip relaxes, the pebble will enter the zone where rocks break rather than bend, and where the mineral veins that sustained generations of Welsh miners were created. It is time to visit the original metal factory.

Breaking the surface

THE PLUMBING OF ROCK

The pebble is a small but perfectly integrated part of a metal factory. This factory has produced copper, silver, zinc, lead and gold (real gold, not its iron sulphide facsimile, pyrite). It is about 100 kilometres long and 60 kilometres across, by about 6 kilometres deep. It is called Wales. The metals have sustained, puzzled, frustrated, and finally abandoned many generations of Welsh miners. Many *hundreds* of generations, indeed, for these metals have been sought, avidly, since at least the Bronze Age, more than 3000 years ago, when shafts were dug through solid rock with little more than hand-held antler bone and rounded cobble.

It is no small feat to chase the metal underground, for its path is tortuous, its presence capricious and its surroundings dangerous. The Welsh miners have been celebrated at home in literature and songs, and also in more surprising quarters, as in the Japanese filmmaker Hayao Miyazaki's portrayal of them in *Castle in the Sky* (a children's animé film, perhaps, but deeply serious at core, like everything that Miyazaki has done). So how is a country-sized metal factory created? Tiny fragments of the answer reside within the pebble.

A streak of white crosses the pebble, cutting across both the strata and the tectonic cleavage surfaces. Cutting both these fabrics, it *must* then be younger. Such evidence of what-came-first and what-came-next is at the heart of geology, and has been so since the very beginnings of the science, since before geological time was pinned and measured by the application of atomic clocks and of fossil time-zonations. And for all today's shiny atom-counting machines and well-stocked libraries and museums, this kind of logic is still the first thing the geologist applies when any new and unfamiliar problem comes into view.

But *what* is it in the pebble that is younger? Peer with the hand lens, and the white streak is resolved as a mineral vein: that is, as a mass of tiny crystals that have grown within a fracture in the rock. It is a common phenomenon, so common that even for an apprentice geologist the identification is soon made at first glance—as a vein, that is, for the white mineral filling the fracture may be quartz, or calcite, or barite, or gypsum. These minerals can look quite alarmingly similar, and so demand a closer look. The mineral in the pebble vein has a glassy lustre and curved fracture-planes, isn't particularly heavy, doesn't fizz with acid, and can't be scratched with a steel blade—so it is most likely quartz.

To seek its origin, we have to move some millions of years after the peak of heat and pressure. The pebble-form has moved higher, as mountains have been eroded away from above it, and as the whole crust has moved upwards in response.

190

FIGURE 9 Large quartz vein in rock.

This phenomenon is called isostasy, and is a grander-scale example of what happens, say, when the centre of gravity of an iceberg rises in the water as its top melts away. In these higher regions, the rock is becoming cooler (dropping once more towards 100°C) and is therefore, more brittle, more prone to fracture—rather than to bend—in response to the strains around it.

These fractures are filled with fluid expelled from the hotter realms beneath. The fluid, as it circulates through the immense fracture-filled network, is constantly in contact with the rock that makes the walls of the fractures. It dissolves certain elements from it—silica, but also a small proportion of the tiny amounts of copper and lead and zinc (gold and silver too) that are present in that huge mass of mudrock. As the water ascends to higher levels, and cools, it can no longer keep so much

191

material in solution, so this precipitates out in those fracture-cavities, filling them with quartz—and here and there, with masses of metal ore.

The metals typically pass through a kind of relay system before they become ores in the vein. They mainly travel underground as metal chlorides, a form in which they are reasonably soluble. When they precipitate, though, they usually do so as sulphides. In the complex plumbing of the subterranean realm, the metal-bearing fluids have met, say, fluids charged with hydrogen sulphide, the sulphur in these in turn having come, perhaps, from the kind of organic-rich laminae that we see in the pebble. The fluids meet, and the metal sulphides are precipitated.

And what lovely things they are! They possess a beauty not seen using the standard microscope that geologists use for looking at thin sections of rocks, for the metal sulphides are opaque, and just appear black when thus viewed. But if one shines light from above, rather then from below, then one sees in the reflections the bright and warm colours of pyrite and chalcopyrite, of galena and sphalerite (Plate 4C-E) and—if one is very lucky—the unmistakable yellow of good Welsh gold. It is a tiny mineral garden, but one with the most exotic of blooms.

Look closely at the tiny vein in the pebble, and it may or may not have a speck or two of some metal ore within it. But the vein will, more than likely, hold a richer treasure also: a memory of the temperature at which it was formed, held within microscopically small samples of fossil water and gas—the very stuff of the fluids from which the vein mineral formed. From them one can recreate, in the laboratory, something of the conditions within that long-disappeared underground rock fracture. Prepare, carefully, a flat sliver of the quartz vein, a little thicker than the normal thin section of rock used in analysis. Take it to a specially designed microscope that has a built-in heater, and examine the sliver at high magnifications. The quartz will be seen to have, imprisoned within it, scattered bubbles of liquid, and within each liquid droplet is a little bubble of gas.

Turn up the heater on the microscope stage, and watch. At a certain temperature, the gas bubble disappears as it dissolves back into the liquid. That represents, fairly well, the temperature at which the original crystal grew, within its fracture deep within the rock, when it trapped that tiny portion of the fluid that bathed and nourished it as it formed.

These days, though, one can do more. One can zap the gas bubble with a laser, and whisk the vaporised elements into a mass spectrometer, to examine the composition of the fluid: its concentrations of carbon dioxide, of sodium and potassium (normally, the fluid is a rich brine) and of other elements. More analysis can be done by closely examining the shape and the chemistry of the crystals in the vein-fill, to establish the history of the fracture as it filled with mineral.

That process may have been anything but slow and steady— indeed, it may have been dramatic, rivalling some of the celebrated fictional depictions of the underground Earth. Perhaps not the centre of the Earth as envisioned by Jules Verne, with giant, mysteriously lit caverns that contain breathable air, and oceans and islands, and dinosaurs too, that our heroes did battle with. That magnificent concept is, alas, impossible (whatever the redoubtable Professor Hardwigg might have affirmed to the contrary). The underground realm is pervasively fluid-soaked (deep working mines have to be continually pumped dry), is regrettably dinosaur-free, and is not penetrable by explorers, no matter how intrepid.

Now and then, one may catch a glimmer of the workings of a quartz vein. Some quartz veins, *very* carefully analysed, have been shown to contain double-pointed crystals, ones that must have grown while suspended in fluid. Quartz crystals could only have been suspended in fluid while they were growing if that fluid was flowing upwards at fast enough speeds (up to a metre a second) to support them, like balls balanced on top of a jet of water.

These underground fracture systems were thus not passive systems, slowly filling with fluid. Rather, they were like a Heath Robinson plumbing system, with makeshift valves and blocks where pressure built up to the state at which it could be explosively released, in a rush of boiling liquid, the pressure drop aiding rapid—almost 'instant'—crystallization. Some systems of this sort can be quite well regulated—witness the punctuality of the Old Faithful geyser in the Yellowstone National Park (which represents the upper end of one of these systems, albeit one that is volcanically heated).

Like any old-fashioned plumbing system, it's a long-lived one, and also one that works in fits and starts. In truth, it started a long time ago, when the mud was first being deposited and was compacting under its own weight. But the distinctive quartz veins, and the metal lodes that they include, mostly date from this later part of its history.

LOW FIDELITY

The pebble, from now on, will still experience events, changes, though few will leave their mark on it. For instance, it will be shaken, periodically, by earthquakes—in its time underground very many thousand tremors will reverberate through it, that were generated both near and far. Certainly, while the mountains were rising, it would have been in a seismically active zone, and some of the locally triggered earthquakes would have been as powerful as any that now shake the Andes or the Himalayas. But, this shaking would have left little mark, no more than do the walls of any football stadium contain a record of the roars of the crowds that have passed through it.

That is not quite true. Just next to the fault line where earthquakes are generated, the rocks may be scratched or jointed or shattered. And, when earthquakes pass through rocks that are brittle but poorly solidified, a little like meringue, they can produce sets of mud-lined fractures. But it's a poor record of the hum and boom of the Earth as it responded to the tremors that passed through it—very low fidelity, for sure. It is humans who invented a means of fossilizing in detail the passing of pressure waves, through the invention of the old shellac 78 rpm record, and then the tape recorder, and now CDs—and there's the less emotionally appealing but more geologically apt seismograph, of course. It is a genuinely and conceptually novel type of fossilization, that has not had any comparable precursors this last 4.5 billion years, not in this Solar System at least. Now there's a thought to bear in mind as you relax and listen to Callas or Caruso.

These earthquakes (which our pebble experienced, but is now mute about) would have come in flurries. The first wave, naturally, with the mountain-building that raised the Welsh Mountains, and likely more as the pressures relaxed, a few million years later, and blocks of crust foundered between others that were left still raised. Then some 50 million years later, more earthquakes would have been triggered as the pebble was probably wrenched upwards again when Wales was caught in another tectonic vice, as another portion of the ocean to the south disappeared into the deep Earth. One cannot glean anything of this from the pebble: case-hardened by its earlier vicissitudes, it could take little direct imprint from these new pressures—or at least none that we have yet had the wit to find. Nor could it keep any tangible record, nearly 200 million years later, when the crust responded to the building of the Alps on one distant side and the opening of the Atlantic Ocean on the other by localized creaks and groans and shifts. A few tens of miles to the north, for instance, a part of what is now the Irish Sea subsided by over three kilometres along a major fault line, and filled with muds much younger than those of the pebble.

The pebble-form would have been almost imperceptibly stirred, too, by the tremors from powerful earthquakes on the other side of the world—just as a seismograph in London or New York today can pick up a signal from a large earthquake from, say, the Pacific Ocean. Now and again there would have been a more energetic shaking from a different source, as, every few million years, a large meteorite impacted somewhere on the Earth. The meteorite that struck Mexico, and that is suspected of killing off the dinosaurs 65 million years ago, would have caused—whether guilty or not of that crime—the entire Earth to ring like a bell. Even 1000 kilometres away from the impact site, it would have convulsed the Earth's structure to the equivalent of about 13 on the Richter scale—many thousands of times more powerful than even a giant terrestrial earthquake. The pebble-form, while still firmly part of its underground rock stratum would have trembled too as the impact waves passed through. If one could, yet one more time, reach back through those abysses of time and space to the pebble-form as it was then, one would surely be tempted to insert the sensor of a seismograph (with *very* long-lived batteries) against it, to catch those passing vibrations.

Most of the other history of that near-half-billion-year span, would simply have passed the pebble by, in its long sojourn underground. The world of the surface, a mere few kilometres above it, might as well have been a million miles away. That world changed beyond recognition, from what was mostly a barren landscape, to one that soon teemed with life. As the pebble was being buried into regions deep enough and hot enough to distil its oil out of it, heavily armoured fish were swimming clumsily high above, colonizing those Avalonian rivers. A little later (geologically) the land greened as plants spread across it. Not much more than 50 million years after the pebble began its underground imprisonment, the luxuriant vegetation of the Carboniferous coal swamps was thriving around the Welsh mountains. When the swamps dried and the plants died, to be replaced by the arid deserts of the Permian and Triassic periods, the pebble was entirely indifferent to this transformation. Nearly all of the Earth's species succumbed to the catastrophe (as yet mysterious, but probably involving mass suffocation on land and in the sea) at the end of the Permian, some 250 million years ago. This dramatically refashioned all of life on Earth (which evolved from the scraps that remained), but the pebble was, likewise, supremely unaffected— as it was when the returning seas of the Jurassic Period lapped at the shores around the Welsh mountains, while dinosaurs— doubtless including the original red dragon of that emerging nation—paced those hills.

HOMEWARD BOUND

At some point, though, the pebble, even while still deep underground, would begin to 'feel' the Earth's surface. Its parent stratum was inching closer, becoming a few centimetres nearer every millennium, as the mountains above were worn away. As the rock mass that lay above diminished, so did the pressure it exerted upon the rocks below. These rocks, no longer squashed tight, could relax, and almost imperceptibly expand upwards. *Almost* imperceptibly. Even a small volume difference is enough to open up minuscule fractures—joints—in the rock. And these fractures extend upwards, as they ascend into rocks that have even less pressure on them, up and up, until eventually they link with the heavily jointed rocks at the surface.

The joints are a pathway for fluids: no longer for ascending fluids squeezed out of the rock itself, but for those that percolate downwards, being ultimately derived from the rain that falls on those eroding mountains. That water brings with it messengers from the surface. Oxygen, for instance, dissolved in the rainwater—though that would generally be used up underground before reaching the deepest recesses. But there would be other chemistry—dissolved carbon dioxide from the surface, say—not to mention the simple power of water as a solvent, acting on the surfaces that it flows along.

And life would return too, after its absence for hundreds of millions of years. The microbes would, at some point, come back to dwell—if not quite yet in the pebble-stuff itself, at least in fractures and nooks and crannies, perhaps just a few thousands of a millimetre across, perhaps just centimetres away. They would not lead brisk lives, these microbes—like the ones that persisted longest on the downwards journey—these would be ones with a metabolism slow and patient enough to survive on the thin pickings at these deep underground levels.

The pebble would feel other changes too. It would feel the sun's heat. Not directly, as in being sun-baked on a shore, not yet; we still have millions of years to wait for that. But up until now it has been heated mainly from the Earth's interior, and the passing of the seasons above has not been felt in that evenness and constancy of that enveloping warmth. And it will take a little while, yet, for seasonal changes to penetrate through to it, and for it to feel the weather.

Weather, though, is not the same as climate. The greater, longer-paced changes of the Earth's climate will reach the underground pebble, well before the rapid successions of summer warmth and winter frosts do. The pebble is perhaps now, some millions of years in the past, several hundreds of metres below the surface. The pebble would cool as it neared the surface, and nearer the surface the circulating water would help draw the Earth's heat away from it. And then the Earth's surface heat—or chill—would find its way down to it.

The Earth's climate was changing, then. Those changes filtered down through to the pebble—and eventually they would go on to shape that pebble, carve it out of its enclosing stratum. When the pebble-stuff was forming as grains on that Silurian sea floor, that sea floor was far south of the equator. And now, when it was nearing the surface again, it was far north. The changes it was to encounter, too, were northern changes—and these were latecomer changes. Had the pebble been formed in a different part of the world, then the crust on which it formed might, by the whim of those deep-Earth convecting currents of almost-solid rock, have been carried south and not north. And then the climate changes would have hit it earlier.

In this present refrigeration of the Earth, ice first grew on Antarctica, over 30 million years ago, and therefore much of the southern hemisphere has been chilled since then. Here in the north, large-scale ice came late, less than 3 million years ago, and this marked the start of the global, bipolar glaciation that defines the world that we currently live in. As the great Laurentide ice sheet grew (via, it is thought, the 'snowgun' effect of water vapour from a seasonally warm north Pacific sea being blown across a cold North American continent), and as those of Greenland and Scandinavia grew, so the chill spread out and ice formed on the hills of Scotland, the Pennines—and Wales.

For the pebble, this would have produced an additional weight (of several hundred metres of ice) that would have halted its upward motion and for a while forced it, and the crust in which it was embedded, downwards. But if little ice grows, and it is just very cold, then the ground (or rather any water in it) simply freezes iron-hard, and that freezing can ultimately penetrate to depths of tens then hundreds of metres, as in the great permafrost terrains of today in Siberia and Canada. The billions of tons of permafrosted rock, insulated by the soil and vegetation on top, take many thousands of years to form, and many thousands to thaw once the climate has warmed. Only the upper surface—the top metre or so that is the 'active layer'—thaws briefly every spring and summer, only to freeze again through the long autumn and winter.

Signs of vanished permafrost, from the glacial phases of the Ice Age, are common in the British landscape, as disturbed soils (the fossilized remains of the active layer), circular hollows where giant ice blisters once grew, distinctive patterns of fractures that formed as the ground froze and, contracting, split apart. The pebble stratum was almost certainly permafrosted as it lay near the surface in the last million years or so—perhaps quite a few times, for there were many glacial phases in the Ice Age. And, given that permafrost is slow to form and slow to thaw, the state of the pebble (deep-frozen or warm) would not have directly related to whether the climate above was bitterly cold or pleasantly temperate. Thus, pulses of warmth and cold going through the underground rock mass would have slowly followed the temperature of the skies above, with a delay of many years. Indeed, one can (with a bit of ingenuity) turn this on its head, and exploit the delay, measuring the temperature structure of the ground in boreholes and reconstructing the history of climate from the way that the waves of cold and warm have passed through (and are still passing through) the ground.

Thus, there is a kind of slow-moving weather underground, of a sort, to complement the climate changes above. Now the pebble is so near the surface, its rise and eventual appearance is affected by that climate. Its rise, indeed, was probably speeded by the cold and ice of the passing glaciations. For although ice is heavy and presses the pliable crust downwards, it is also abrasive, and wears away at the mountains that form part of that crust. As those mountains are converted to debris, and carried away to lower ground as boulder clay and as sands and gravels, so the crust rises higher to compensate, and so the pebble stratum speeds its progress towards the surface and its own appointment with the erosive powers of wind and rain.

The cold begins the process of erosion, while the rocks are still underground. In forming permafrost, the water underground turns to ice. Ice is less dense than water—a simple observation familiar to anyone who has left an empty milk bottle out in the rain and found that a hard frost has come overnight. Simple—but water is a definite oddity in this respect, for most materials become denser and more compact when they freeze. It is lucky enough for the history of life, though, for the oceans would have been much more inhospitable places if ice sank rather then floated. In such a world, it would have formed a thick layer, prevented from melting by the insulation of the unfrozen water above, pressing down on the sea floor and growing to fill most of the ocean basins. In such alternative-universe seas, which might resemble an undersea Antarctica, microbial life would doubtless have found a way to cope—but it would have been no place to evolve lobsters or seashells or dolphins.

A quirk of chemistry has allowed a complex tapestry of multicellular life to evolve—but here we are simply interested in a much simpler thing: its destructive properties. As the water freezes in the underground cracks and cavities and minute spaces between grains, it expands, with enormous force (far more than is necessary to fracture a humble milk bottle). Existing fractures are forced wider apart, and new ones form, to fill with percolating water in turn, when the thaw comes. At depth in permafrost, the freezing and thawing follow cycles lasting many thousands of years. It is only much nearer the surface that the rock reacts more quickly to the passing episodes of warmth and cold, and freezing and melting succeed each other more quickly and more frequently. It is an effective pre-fracturing, and the patterns of breakage will determine the size and shape of the blocks to emerge at the surface. And, when those are broken down further, they also influence the size and shape, therefore, of the pebbles that are to be derived from them.

We are not quite there yet. The landscape needs to form the shore, along which the pebble is to be washed by the waves. A shoreline is the boundary between the land and the sea, and so a boundary that is mobile, ephemeral—particularly during the crazy climate oscillations that are characteristic of a partly ice-bound world. As ice grows, water is taken out of the sea. So land grows, the oceans shrink, and the world's shorelines move outwards. As ice melts, thousands of years later, the waters flood back into the expanding oceans, and so shorelines move back inwards again.

So, 20,000 years ago, the pebble lay not far underground, under a thin carapace of rock that itself was covered by glacier ice. Today, this lies on the coastline, but at that time sea level was more than a hundred metres lower than now, so the seashore lay out to the west. Perhaps, though, not as far out as you might expect for this scale of sea level fall, for the ice was also pressing down on the Welsh hills, pushing them downwards by some tens of metres.

Nevertheless, a few thousand years later, ice around the world was melting, sometimes catastrophically, and sea levels were rising. This was the transition towards the familiar world of today, which we call home. Here, around the spot of the soon-to-be-released pebble, the sea was flooding in—but the land was also rising (rather more slowly) as it rebounded isostatically with the disappearance of the weight of ice. So the coastline was constantly changing, migrating as the rising land vied with incoming sea.

About 5000 years ago, sea level had more or less stabilized at the position it is in today, and the modern coastline started to form. Humans had arrived too—or rather returned, for they had been present here earlier, 100,000 years ago and more, in previous warm interludes of Ice Age climate. They hunted, they began to farm (a novelty, this, for humans)—and went in search of the shining metal in the veins in the rock. The pebble was still a pebble-to-be, within its pebble stratum on that hillside, into which the waves of the newly adjusted sea were biting. But as the waves bit deeper into the rock, it was only a few millennia away from assuming its unique, individual status, its form as a pebble.

INTO THE LIGHT

The eventual excavation of the pebble—perhaps only a few centuries ago—was brutally efficient. The waves, breaking on the rocks of the newly formed cliff, drove air into the fractures and cavities of the rock, a rock that had been weakened by the growth and then the thaw of long-vanished ice. Those fractures and cavities in turn reflect far more distant events—the formation of the stratal layers on the Silurian sea floor, and then their crumpling, and the refashioning of their texture to form a tectonic cleavage (along which the rock now splits into slabs)—and then the fracturing that formed as the mountains rose. Compressed air, with tons of storm-driven water behind it, acts like a chisel. Applied thousands of times over many storms (and aided and abetted by the boulders also hurled against the cliff by the waves), it eventually loosens slabs of slate, and then finally levers them out—sometimes dramatically, when sections of the cliff collapse.

One of those slabs is the parent of the pebble, released, after a little over 420 million years, back to the surface. Perhaps only a few centimetres thick (though it might be as much as a metre across) its smooth surfaces are defined by the tectonic cleavage, imposed on the strata about 396 million years ago, it lies at the foot of the cliff. It contains within it our pebble, and quite a few others too. To release them—or more precisely to shape them—one needs first to smash the slab.

The storm waves duly oblige, picking up the slab and hurling it against the cliff until it breaks. The now-nearly-pebble breaks off, as a sharp-edged shard of the slab. It has all the features of its own narrative, that we have followed until now—the stripes of pale and dark rock, the fossils both large and small (or, from a human perspective, small and microscopic), the mineral menagerie and the interwoven chemical and isotopic patterns—but it needs shaping.

The waves can do that too, now washing the more easily transportable fragment along the shingle of the beach, the swash and the backwash constantly jostling it among its countless fellow pebbles along that shoreline. As it travels, pushed along by longshore drift, the thousands of collisions wear it away, smoothing the sharp edges, creating the rounded outline that is just waiting to be picked up—but perhaps not just yet.

For there are another four chronometers to add to the fine tally that the pebble has already acquired. Before we add these latecomer timepieces, let us take stock of the ones already imprinted into the fabric of the pebble. There is the pattern of the neodymium isotopes, which tell us when the stuff of the pebble was released from the Earth's mantle, and when Avalonia was formed: that's one. And then there are the zircon grains, crystallized in magma chambers and deep mountain roots perhaps a billion and more years ago: two. Those sparse rhenium and osmium atoms subsequently bore witness to the falling of the mud flakes on to the sea floor: three. Then, the crystallization of monazite, and the reshuffling of rubidium, deep in the Earth, perhaps as oil was formed and clay minerals transformed: four and five. And after that there was the decay of radioactive potassium in the mica crystals that grew around fossils that were being squeezed by the pressures of mountain-building—that makes six.

This is a conservative tally, for in abbreviating the story of the pebble so that it can fit into one small book and not into several weighty tomes, a few chronometers have been missed out (there are ways of dating mineral veins, for instance)—not to mention those hidden in the pebble that we have yet to discover—there will be some of those, undoubtedly. So, with the four newcomers we will have ten in total, at least (it is nice to get into double figures).

Geologists need these chronometers, for they have to deal with colossal amounts of time. To have evidence of things that have happened without knowing when and in what order they happened is a recipe for profound confusion—a kind of a soup of events, well stirred and utterly incomprehensible. Therefore, they have racked their brains to find as many ways as possible to say *what* happened *when*. As racking goes, it has been quite fruitful, all told. And as geology deals with the only-just-yesterday as well as with the deep past, there are other questions that can be asked. For instance, how long has the pebble been a pebble?

Let us allow the pebble to be caught up in another great storm, a 1000-year storm. It is tossed high onto a shingle bank at the back of the beach—so high and so far that normal storms cannot dislodge it to bring it back, by backwash, to the active, continually moving pebble streams of the wave-washed beach. It lies still on the very top of that far shingle bank, exposed to the elements. And three clocks start to tick, while a fourth is due to be activated shortly—in 1945, to be precise.

Now, one clock starts forming on the upper surface, another on its lower one, yet another in one spot—or perhaps two—on its surface, while the means of constructing the fourth one is germinating in human minds. Let us take the upper clock first, as it has a grandeur about it that commands our first attention.

From the sky, a bombardment is taking place. Cosmic rays are constantly speeding across space, and smash into anything that gets into their way. They are smashing into you and me, and also into the exposed pebble surface where, over time, they do significant damage. When they collide with silicon atoms, the atoms are broken up, with one part of the debris being a short-lived, highly radioactive form of aluminium, while oxygen, similarly mistreated, is transformed into radioactive beryllium. If the bombardment goes on for long enough (a few millennia) then enough of these radioactive by-products accumulate to measure by the marvellous atom-counting machines of today; thus, counting the number of the new atoms gives an estimate of how long that rock surface has been exposed. This is cosmogenic dating. It is a baby technique (not much more than two decades old), but in that short time it has become a standard means of dating landscapes and rock surfaces.

On the underside of the pebble, some atoms are quietly acquiring their own more local damage. The pebble, while it was being washed along the beach, being turned over and over by the waves, was being sunbathed for the first time for hundreds of millions of years. The sun's rays not only warmed the pebble surface: they repaired radiation damage done to the molecular lattices of some of the minerals, notably quartz. This accumulated damage was caused by radiation generated from within the pebble, by its own natural radioactivity. Hide a pebble surface from the sun, and that self-healing mechanism is switched off. The crystal lattices, over centuries and millennia, once more acquire minuscule defects and distortions. Shine a controlled light on them in the laboratory, and the lattices spring back into shape, giving off a tiny flash of radiation as they do so. The size of this flash, carefully measured, is a measure of how long those crystals have been in the dark. Optically stimulated luminescence, it's called, and it's used so often by scientists who study the Ice Ages that it has its own acronym, OSL. Very useful it is too, in providing a guide to how long sediment has been buried.

And then there's the most natural, the most *ecological* of dating methods for a pebble surface. Above that exposed pebble surface, high on the shingle bank, spores will be drifting. Some of them will land on the surface. Life will begin to colonize the pebble, getting a grip even on that smooth surface. Lichen—that strange amalgam (symbiosis, more precisely) of alga and fungus—has begun to grow. Lichen grows extremely slowly, at perhaps at a millimetre a year, and steadily, and thus the size of a lichen patch gives an indication of how old an exposed rock surface is. This technique even has its own name—lichenometry. There are ifs and buts, of course. Lichen will probably not begin to grow straight away on a surface; and environmental factors can affect lichen growth—pollution, for instance. But nevertheless it is refreshing to have a dating technique that one can use with a magnifying glass and ruler, rather than with several hundred thousand pounds' worth of *very* delicate equipment.

Then there is another clock, one of our very own making. It started ticking at forty-five seconds and twenty-nine minutes after five o'clock in the morning (local time) in Alamogordo, New Mexico on 16th July, 1945, when the first atomic bomb was tested. That, and the succeeding Hiroshima and Nagasaki bombings, that killed some 220,000 people in total, and the succeeding above-ground atomic weapons tests (before they were banned)—and, more recently, the Chernobyl nuclear accident—produced new, artificial radionuclides that spread around the world. These radionuclides, including plutonium and a long-lived caesium isotope, can be detected in virtually everything that has been at the surface since that time. There will be some of these new radionuclides, for certain, on the pebble surface—the nutrient-hungry lichens will have been especially effective at scavenging them from the air and from the raindrops.

What a diversity of time, and of events! Now, at last, we come to our brief encounter with the pebble, before we send it on its way into the future. Another mighty storm, and it is dislodged, back to the beach. Washed back and forth once more, it quickly loses its lichen coating (and therefore that particular clock), and the optically stimulated clock is reset too. The cosmic clock, though keeps ticking, as the rays keep smashing in from outer space—oh, and there's no escaping the man-made radionuclide clock, either, anywhere at the Earth's surface.

The pebble, now shiny and patterned, catches one's eye. One can pick it up and sit, for a while, and consider it. It's a nice afternoon. It is tempting to sit on the beach and muse on its history a little, while the Sun is still high in the sky, and one can make oneself comfortable. Time passes.

It's getting dark. Time to go home. The pebble is tossed aside—there are many more, after all. And it still has a destiny. Many destinies, in fact.

Futures

BREAKING UP

The pebble is on the beach, once more, unmarked by its brief contact with human sentience. Almost unmarked. The fingerprints that it lightly bears will, however, be washed away by the next tide. It has a long future, still, but probably not as a pebble—though quite how long it remains as a pebble may well depend on human action. Not on immediate, direct human action—whether it is scooped up by a digger and converted into concrete for a sea-front esplanade, for instance, or even collected as a souvenir by some passing tourist. Either of these fates should cause only a brief deflection from its long-term future (the esplanade is, after all, only a cliff to be attacked by the elements, while beach souvenirs are soon discarded). A larger perturbation of its trajectory more probably hinges on wider human effects—but more of that anon. We might assume, first, that nature runs its course.

A pebble on a beach, its natural environment, is changing all the time. Not long ago, it was part of a slab of slate in a cliff, then it briefly became an angular chunk of rock, before the waves and water smoothed it down. They are still smoothing it, wearing away at it, making it smaller. Even the contact with human hands probably removed a grain or two. A pebble has the appearance of permanence, but it is not permanent. How long does it take to wear down a pebble?

This can happen astonishingly quickly. Even over a single tide, being washed backwards and forwards by every incoming wave, a pebble can become detectably lighter—by less than one tenth of one per cent, admittedly, but that weight difference can easily be measured using modern electronic scales. Over a season, on an exposed part of the coast, a pebble can lose between a third and a half of its mass. The rates will vary—on a stormy day the banging of pebbles against each other can produce distinct percussion marks on their surfaces, while on a calm day the attrition rate will drop markedly.

Night and day, though, the pebble is disintegrating. What can save it? Well, perhaps we can, temporarily, though at a high cost to ourselves. Not so much by taking it off the beach and putting it into a drawing-room cabinet, or a museum. But a geologically proven way to stop the erosional processes on a beach is to drown it. And it seems more likely than not, as things now stand, that one by-product of our civilization will be a geologically sudden sea-level rise, of some few metres over the next few centuries.

The shingle beach, and the cliff, will be taken below the destructive wave zone, and covered with mud and silt. If the remnant of our pebble is still somewhere on that beach, it may be smuggled across many millennia still, in what is left of its pebble-form, for it may take a long time for the waters to recede. But recede they eventually will, in perhaps 100,000 years from now. The pebble will be exhumed, and put back into the mill of erosion.

Whether now, or whether in that post-industrial future, the pebble will be dismantled. Its components, that have stayed together for so long, will now part company, and go their separate ways. Some will go on, in a very short space of time, to literally encircle the world, by air and by sea. Other components will rest more locally. But as time passes, even these will diffuse outwards, as the parts of the pebble-that-was are scattered far and wide.

The travel paths are to some degree predictable, at least initially. The quartz grains in the pebble will be released into the sand and silt on the beach. They will not be quite as they were, though, when they initially arrived on that Silurian sea floor. They have been reshaped in their long contact with those underground fluids as the strata were first compacted, and then crumpled as the Welsh mountains were built. The original grains have been partly dissolved, partly cemented over by extra silica, partly fused together by the effects of pressure solution. So what emerges will be irregular clusters of former grains, or parts of them, and these will join the other grains on the beach, and be swept along the coast by longshore drift, or washed out to sea, into deeper water by storms and tides, being separated all the time as the lighter grains are driven farther and faster, and the larger ones trundle along more slowly.

Some grains, though, will be pretty much in the shape and form that they were when they arrived, hundreds of millions of years ago, on that Silurian sea floor. These are the well-nigh indestructible zircon grains and their kin, such as rutile and tourmaline. They are physically tough and chemically resistant, and the pressures of mountain building and corrosive effects of underground fluids will have had little effect on them. Released from the pebble again by erosion, they may initially be encased in some silica cement, say, or have a quartz grain or mica flake stuck to them. But those will soon be abraded off, as the grains travel among the sand grains, jostling and tumbling along. For them it may be their second or third or fourth such journey in one or two billion years, being recycled from one set of strata into another.

These resistant grains are heavy. Their density means that they separate out from the quartz grains, and can be winnowed, concentrated into parts of the sea floor where the currents are strongest, or accumulate in depressions and potholes (rather as do gold grains in, say, the alluvial deposits of the mighty Yukon and Klondike rivers, where prospectors have to think through the effects of current velocity and shear and drag to find the best places where they now lie, to stake the best claim).

The micas that make up the bulk of the rock will flake away, along those perfect book-like mineral cleavage planes. Light and delicate, these are washed (or blown) away easily. They have gone back to being mud, and separate quickly from the sand grains, to join other mud particles travelling along that shoreline. These particles can now travel long distances suspended in water, to eventually settle on a mud flat or salt marsh in some sheltered estuary, or to be carried out onto a far and deep sea floor. Once settled, they can then stick as a cohesive layer, perhaps later to be re-eroded, ripped up as lumps, disaggregated, and carried further.

These particles are now firmly back in the land of the living. In the shallow seas, their contact with the living world will be not dissimilar to that in their transect across the shallow Silurian shoreline. They will be swallowed whole by mud-eating worms, filtered out by the delicate fan-like apparatuses of filter-feeding organisms, pushed aside by crawling and walking crustaceans and, everywhere, covered in those omni-present microbes. Particles blown onto land, though, will encounter a landscape transformed. They will land in thick, humus-rich living soils, penetrated almost everywhere by the roots of plants—a far cry from those largely barren Silurian landscapes.

The microbes will, in part, be greeting the arrival of the remnants of organic matter still left in that slate, trying to make a meal of it as it is released, as carbon wisps, from the pebble. This will be often unsuccessful, for much of the carbon is now little more than graphite and is pretty well indigestible, even for microbes. Those black graphitic wisps are simply washed away with the mud flakes, to form part of a new sediment layer and ultimately a new stratum, reburying that carbon before it rejoins the cycle of living organisms.

But some of that carbon will be digested, consumed and join in the great cycle of life, ascending the food chain from microbe to protozoan to worm to fish—to human, perhaps. Metabolized, excreted, reassimilated, it will be travelling all the while, carried by both ocean waters and by the moving and migrating bodies of plants—those mobile planktonic algae—and animals. Respired, it becomes a gas, carbon dioxide, first dissolved in those ocean waters, and then released to the atmosphere where it travels with the winds, circumnavigating the world; it will be redissolved in rainwater and taken back to land or sea, to corrode rock formations or the shells of plankton; or be taken up by plants on land or sea, which are in turn eaten, before the carcasses of herbivores and predators alike fall to the sea floor, to be buried in some newly forming stratum as a prelude to deep Earth burial once more.

At this stage, almost every atom of carbon soon follows its own separate path from those that had been its near neighbours underground for so long, as this component of the pebble is dissipated finely across the globe. Each has its own fate, and each from now will pursue its own path, unlikely ever to rejoin its former neighbours. Some may even be carried into space, venturing into the stratosphere and being stripped away from the Earth by the solar wind; others will be carried deep into the Earth, caught up on some descending oceanic plate and be carried on down into the mantle—there perhaps to be incorporated into a growing diamond crystal. It is a diaspora without compare.

Other pebble components are long-distance travellers too. The pyrite that infills the modular homes of the graptolites, and forms the many framboids scattered within the rock of the pebble, survived the pressures of mountain building, but cannot long tolerate the mild sea breezes and rain showers of the Welsh coast, or the immersion in sea water. This golden iron mineral quickly tarnishes, the sulphide oxidizing to sulphate, while the iron becomes a hydroxide. The graptolite is no longer filled with gleaming fool's gold but with friable orange rust; often even this falls out, leaving an empty space—the same situation, effectively, as when the dead colony lay on the Silurian sea floor. The sulphate released from the pyrite may link with calcium to form tiny translucent crystals of selenite, a form of gypsum—or it may simply join the huge reservoir of dissolved sulphate in the sea.

Once dissolved in the seas, a sulphate ion may simply stay there for thousands of years, travelling the ocean currents. Its eventual fate may be to drift near to the sea floor and diffuse into the surface sediment layer, there to cross into the anoxic zone, and be used as an energy source by a sulphate-reducing microbe, to be converted into sulphide and once more into pyrite. Or, the sulphate may drift into a shallow lagoon on some hot and arid coastline and be crystallized as gypsum when that water evaporates. Or, perhaps, it might be assimilated by marine algae and released as an aerosol of dimethyl sulphate into the atmosphere above, and in this form it will 'seed' minute droplets of water, to make clouds and rain. It is another grand parting of the ways for sulphur atoms that had shared the same underground home (albeit one that originally belonged to the graptolites) for so long.

Other atoms of the pebble are harder to part. The monazite crystals that grew while—and perhaps because—oil was stewing from those rocks are not quite as tough as the zircons, not least because of their cargo of clay impurities, but they are resistant enough. They typically erode out of slate as elliptical grains about as large as the head of a pin. Like the zircon crystals, they are dense, and so will concentrate wherever the ordinary, lighter sediment grains are winnowed away. Here and there, in Welsh streams and rivers, such eroded crystals have been found in abundance as 'monazite sands'. How far, though, will they ultimately travel? This is an open question. As distinct phenomena, they have been only recently discovered. Have they been reworked, like the zircons, into younger strata? No-one has yet looked.

The particles that once travelled immense distances, in space and on Earth, to eventually meet in the pebble, are now separated. They will move farther and farther apart. Most will remain on Earth, rather than travelling outwards (like, perhaps, those few stray carbon atoms) into the cosmos. Almost all will—sooner or later—be incorporated into new strata that will be buried in turn, be compacted, hardened, mineralized, crumpled in mountain belts, uplifted and eroded. And then countless new pebbles will arise, in different places and at different points in time, each carrying within it a minuscule part of our original pebble. Many of these will still be made up of sand or silt or mud, but particles of our pebble will turn up in limestones, also, and salt deposits, and new deposits of oil and gas, and in magmas too, to make up a submicroscopic component of some basalt or granite. The new strata will carry within them the remains of very different animals and planets, as the Earth's freight of living organisms carries on evolving, in Darwin's words, endless forms most beautiful and most wonderful.

EPILOGUE

How long will these cycles continue? Unless some unforeseen catastrophe strikes, natural or man-made, they will last for another billion years, perhaps two. It is enough to make a few more generations of such pebbles. Then, when the dying Sun grows larger, and its thermonuclear fires begin to burn brighter, the Earth will lose its oceans, as they boil away and are stripped off into space, and its oxygen-rich atmosphere, and its life, the microbes last of all. Life, in any event, may have fared poorly for some time. The Earth's metal core will have frozen, or mostly so, and so the Earth will no longer have a magnetic field, or the protection that it affords from the solar wind or cosmic rays.

Our pebble will then be scattered far across the Earth, within strata that will now, mostly, be fossilized relics of that earlier, kinder planet. It will be a strange Earth, in this, its old age.

Some kind of gaseous envelope will likely still surround it, so winds of a sort there will be, but there will be no rain, no streams or rivers or lakes. Loose sediment will be wind-whipped into dunes, as happens on Mars today. Cliffs and crags will now and again collapse in rock-avalanches, for gravity will not be any less a force than it is today. Mountain ranges will still form, for a while, until the great engine of plate tectonics shuts down. It is an engine that will be weaker in any event, as the Earth's radioactive heat that drives it declines. The removal of that great planetary lubricant, water, will eventually stop it as effectively as removing the oil from a car motor, in rendering the subduction of ocean plates too friction-ridden to be possible.

The Earth's remaining heat must then still have to escape, so volcanism will continue for some time yet. Our planet, increasingly dimly lit by the dying fires of the Sun, will be entering a perpetual night, illuminated only by the distant stars of this galaxy. In such a scenario, on our far future planet, the pebble cycle will be over, and a sleep eternal will begin. After a solar system has run its course, what remains is ashes.

But it is nice, always, to have an escape clause. Rebirth might come if the dead remains of our Earth are swept, along with Mercury and Venus, into the Sun as it grows hugely in its final red giant phase, perhaps some five billion years from now. It will be touch and go whether it does. The Earth's orbit will then be looping outwards as the Sun loses mass, but it may not move fast enough. The Earth might just be caught up in the enveloping edge of the billowing Sun, be dragged inwards, spiral down and be vaporized, as the outer layers of our dying Sun are blown outwards, while its core collapses into a white dwarf which, at the end, is not much bigger than the Earth itself.

There will not be a supernova to mark its passing—nor even an ordinary nova, to generate phantasmagoric nebulae like those—the Cat's Eye, the Siamese Squid, the Red Spider—captured today by the far-seeing eye of the Hubble Telescope. Our Sun is simply too small for such melodrama. Nevertheless, there will be a kind of mineral diaspora from our Sun's final outburst, as some of that vapour from our once-beautiful planet is swept into interstellar space.

And from there, that cosmic dust, with a few of our pebble atoms in it, would drift across the galaxy. Eventually, it may be swept up into the birth of a new star system, which develops planets of its own. It's a long shot—but not impossible. This is how our story started, after all, on this planet, with this pebble.

So perhaps it can begin again.

FURTHER READING

There are few books that deal with such geologically constrained subject matter. I do have one distinguished, indeed legendary, predecessor. Gideon Mantell, country doctor and the first dinosaur scientist of the modern age, wrote *Thoughts on a Pebble*, published in 1836. It is charming, still informative, and (though out of print a century and more) now easy to access, courtesy of Google. The geology in it is remarkably prescient, and it is graced with poems by Byron, Walter Scott, Percival—and by one Mrs Howitt, whose rhyming couplets portrayed 'with much force and beauty' the pearly Nautilus. For a latter-day book in kindred spirit, I highly recommend *Sand*, by Michael Welland (OUP 2008)—the never-ending story of this inexhaustibly various material, beautifully told.

Ranging more widely than the consideration of sedimentary particles, there are books on the Earth that yielded this pebble, which are both instructive and entertaining. A modest if various selection might include:

Clarkson, E.N.K. & Upton, B. 2009. *Death of an Ocean: A Geological Borders Ballad*. Dunedin Academic Press.

Fortey, Richard. 2005. *Earth*. Harper Perennial.

Hardy, A. 1956. *The Open Sea: Its natural history. Part 1. The world of plankton*. Fontana New Naturalist.

Kunzig, Robert. 2000. *Mapping the Deep: the Extraordinary Story of Ocean Science*, Sort Of Books.

Levi, P. 1975. *The Periodic Table*. Penguin Classics.

Lewis, C.L.E. & Kuell, S.J. (eds) 2001. *The Age of the Earth: From 4004 BC to AD 2002*. Geological Society, London.

Nield, Ted. 2007. *Supercontinent: Ten Billion Years in the Life of Our Planet*. Granta Books.

Osborne, Roger. 1999. *The Floating Egg: Episodes in the Making of Geology.* Pimlico.

Palmer, D. & Rickards, B. (eds) 1991. *Graptolites: writing in the rocks.* Boydell Press, Woodbridge, Suffolk.

Redfern, Martin. *The Earth: A Very Short Introduction.* Oxford University Press.

Rhodes, F.T., Stone, R.O. & Malamud, B.D. 2008. *Language of the Earth* (2nd edition). Blackwell Publishing.

Stow, Dorrik. 2010. *Vanished Ocean: How Tethys Reshaped the World.* Oxford University Press.

Trewin, Nigel. 2008. *Fossils Alive!: New Walks in an Old Field.* Dunedin Academic Press.

Weinberg, S. 1977. *The first three minutes: a modern view of the origin of the Universe.* Basic Books.

BIBLIOGRAPHY

These papers are a selection of the technical literature that provide the science—the observations, the analyses, the deductions—upon which the story of this pebble is based. There is much, much more; the depth and variety of collective human knowledge is a marvel as astonishing as anything in the natural world—and provides a reflection of that world. It is a veiled and imperfect reflection, of course, for there is still much to discover.

CHAPTER 1

Ball, T.K., Davies, J.R., Waters, R.A. & Zalasiewicz, J.A. 1992. Geochemical discrimination of Silurian mudstones according to depositional process and provenance within the Southern Welsh Basin. *Geological Magazine* **129**, 567–572.

Dutch, S.I. 2005. Life (briefly) near a supernova. *Journal of Geoscience Education* **53**, 27–30.

Herbst, W. et al. 2008. Reflected light from sand grains in the terrestrial zone of a protoplanetary disk. *Nature* **452**, 194–197.

Russell, S. 2004. Stars in stones. *Nature* **428**, 903–904.

CHAPTER 2

Jacobsen, S.B. 2003. How old is Planet Earth? *Science* **300**, 1513–1514.

Kiefer, W.S. 2008. Forming the martian great divide. *Nature* **453**, 1191–1192.

Lister, J. 2008. Structuring the inner core. *Nature* **454**, 701–702.

Murphy, J.B., Strachan, R.A., Nance, R.D., Parker, K.D. & Fowler, M.B. 2000. Proto-Avalonia: A 1.2–1.0 Ga tectonothermal event and constraints for the evolution of Rodinia. *Geology* **28**, 1071–1074.

Priem, H.N.A. 1987. Isotopic tales of ancient continents. *Geologie en Mijnbouw* **66**, 275–292.

Widom, E. 2002. Ancient mantle in a modern plume. *Nature* **420**, 281–282.

Witze, A. 2006. The start of the world as we know it. *Nature* **442**, 128–131.

Wood, B.J., Walter, M.J. & Wade, J. 2006. Accretion of the Earth and segregation of its core. *Nature* **441**, 825–833.

CHAPTER 3

Carrapa, B. 2010. Resolving tectonic problems by dating detrital minerals. *Geology* **38**, 191–192.

Merriman, R.J. 2002. The magma-to-mud cycle. *Geology Today* **18**, 67–71.

Morton, A.C., Davies, J.R. & Waters, R.A. 1992. Heavy minerals as a guide to turbidite provenance in the Lower Palaeozoic southern Welsh Basin; a pilot study. *Geological Magazine* **129**, 573–580.

Phillips, E.R. et al. 2003. Detrital Avalonian zircons in the Laurentian Southern Uplands terrain, Scotland. *Geology* **31**, 625–628.

CHAPTER 4

Davies, J.R., Fletcher, C.J.N., Waters, R.A., Wilson, D., Woodhall, D.G. & Zalasiewicz, J.A. 1997. Geology of the country around Llanilar and Rhayader. *Memoir of the British Geological Survey*, Sheets 178 & 179 (England and Wales), xii + 267 pp.

CHAPTER 5

Cave, R. 1979. Sedimentary environments of the basinal Llandovery of mid-Wales. In: Harris, A. L., Holland, C. H. & Leake, B. E., (eds) *Caledonides of the British Isles: Reviewed.* Geological Society, London. Special Publications **8**, 517–526.

Diaz, R.J. & Rosenberg, R. 2008. Spreading dead zones and consequences for marine ecosystems. *Science* **321**, 926–929.

Jones, O.T. 1909. The Hartfell–Valentian succession in the district around Plynlimon and Pont Erwyd (North Cardiganshire). *Quarterly Journal of the Geological Society, London* **65**, 463–537.

Page, A., Zalasiewicz, J.A., Williams, M., & Popov, L.E. 2007. Were transgressive black shales a negative feedback modulating glacioeustasy in the Early Palaeozoic Icehouse? From: Williams, M., Haywood, A.M., Gregory, F.J. & Schmidt, D.N. (eds) *Deep-Time Perspectives on Climate Change: Marrying the Signal from Computer Models and Biological Proxies*. The Micropalaeontological Society, Special Publications. Geological Society, London, 123–156.

Thornton, S.E. 1984. Basin model for hemipelagic sedimentation in a tectonically active continental margin: Santa Barbara Basin, California Continental Borderland. In: Stow, D.A.V. & Piper, D.J.W. (eds) *Fine-Grained Sediments: Deep-Water Processes and Facies*. Geological Society, London, Special Publication **15**, 377–394.

CHAPTER 6

Crowther, P.R. & Rickards, R.B. 1977. Cortical bandages and the graptolite zooid. *Geology and Palaeontology*, **11**, 9–46.

Katija, K. & Dabiri, J.O. 2009. A viscosity-enhanced mechanism for biogenic ocean mixing. *Nature* **460**, 624–626.

Lapworth, C. 1878. The Moffat Series. *Quarterly Journal of the Geological Society, London* **34**, 240–346.

Loydell, D.K. 1992–93. *Upper Aeronian and lower Telychian (Llandovery) graptolites from western mid-Wales*. The Palaeontographical Society, London, Publ. 589 for vol. 146 (1992), 1–55; publ. 592 for vol. 147 (1993), 56–180.

Molyneux, S.G. 1990. Advances and problems in Ordovician palynology of England & Wales. *Journal of the Geological Society, London* **147**, 615–618.

Paris, F. & Nõlvak, J. 1999. Biological interpretation and paleobiodiversity of a cryptic fossil group: the 'chitinozoan animal'. *Geobios* **32**, 315–324.

Rickards, R.B., Hutt, J.E. & Berry, W.B.N. 1977. Evolution of the Silurian and Devonian graptoloids. *Bulletin of the British Museum (Natural History) Geology* **28**, 1–120, pls 1–6.

Rushton, A.W.A. 2001. The use of graptolites in the stratigraphy of the Southern Uplands: Peach's legacy. *Transactions of the Royal Society of Edinburgh: Earth Sciences* **91**, 341–347.

Sudbury, M. 1958. Triangulate monograptids from the *Monograptus gregarius* Zone (lower Llandovery) of the Rheidol Gorge (Cardiganshire). *Philosophical Transactions of the Royal Society of London* B**241**, 485–555.

Underwood, C.J. 1993. The position of graptolites within Lower Palaeozoic planktic ecosystems. *Lethaia* **26**, 189–202.

Zalasiewicz, J.A. 2001. Graptolites as constraints on models of sedimentation across Iapetus: a review. *Proceedings of the Geologists' Association* **112**, 237–251.

Zalasiewicz, J.A., Taylor, L., Rushton, A.W.A., Loydell, D.K., Rickards, R.B. & Williams, M. 2009. Graptolites in British stratigraphy. *Geological Magazine* **146**, 785–850.

CHAPTER 7

Armstrong, H.A. et al. 2009. Black shale deposition in an Upper Ordovician–Silurian permanently stratified, peri-glacial basin, southern Jordan. *Palaeogeography, Palaeoclimatalogy, Palaeoecology* **273**, 368–377.

Bates, D.E.B. & Loydell, D.K. 2003. Parasitism on graptoloid colonies. *Palaeontology* **43**, 1143–1151.

Brenchley, P.J. et al. 1994. Bathymetric and isotopic evidence for a short-lived late Ordovician glaciation in a greenhouse period. *Geology* **22**, 295–298.

Loydell, D.K., Zalasiewicz, J.A. & Cave, R. 1998. Predation on graptolites: new evidence from the Silurian of Wales. *Palaeontology* **41**, 423–427.

Selby, A. & Creaser, R.A. 2005. Direct dating of the Devonian–Mississippian timescale boundary using the Re–Os black shale geochronometer. *Geology* **33**, 545–548.

CHAPTER 8

Chopey-Jones, A., Williams, M. & Zalasiewicz, J.A. 2003. Biostratigraphy, palaeobiogeography and morphology of the Llandovery graptolites *Campograptus lobiferus* (M'Coy) and *Campograptus harpago* (Törnquist). *Scottish Journal of Geology* **39**, 71–85.

Cocks, L.R.M. & Fortey, R.A. 1982. Faunal evidence for oceanic separations in the Paleozoic of Britain. *Journal of the Geological Society, London* **139**, 465–478.

Cocks, L.R.M. & Torsvik, T.H. 2002. Earth geography from 500 to 400 million years ago: A faunal and palaeomagnetic review. *Journal of the Geological Society of London* **159**, 631–644.

Dunlop, D.J. 2007. A more ancient shield. *Nature* **446**, 623–625.

Olsen, P. 2009. Tectonics at the Earth's core. *Nature Geoscience* **2**, 379–380.

Vandenbroucke, T.R.A. et al. 2009. Ground-truthing Late Ordovician climate models using the palaeobiogeography of graptolites. *Palaeoceanography* **24**, PA4202, doi:10.1029/2008PA001720.

Wilson, D., Davies, J.R., Waters, R.A. & Zalasiewicz, J.A. 1992. A fault-controlled depositional model for the Aberystwyth Grits turbidite system. *Geological Magazine* **129**, 595–607.

CHAPTER 9

Bjerreskov, M. 1991. Pyrite in Silurian graptolites from Bornholm, Denmark. *Lethaia* **24**, 351–361.

Nealson, K.H. 2010. Sediment reactions defy dogma. *Nature* **463** 1033–1034.

Raiswell, R. & Berner, R.A. 1985. Pyrite formation in euxinic and semi-euxinic sediments. *American Journal of Science* **285** (8), 710–724.

Smith, R. 1987. *Early diagenetic phosphate cements in a turbidite basin.* Geological Society of London, Special Publications **36**, 141–156.

CHAPTER 10

Evans, J.A., Zalasiewicz, J.A., Fletcher, I., Rasmussen, B. & Pearce, N.G. 2002. Dating diagenetic monazite in mudrocks: constraining the oil window? *Journal of the Geological Society of London* **159**, 619–622.

Evans, J.A., Zalasiewicz, J.A. & Chopey-Jones, A. 2008. The effect of small- and large-scale facies architecture of turbidite mudrocks on the behaviour of isotope systems during diagenesis. *Sedimentology* **56**, 863–872.

Evans, J.A. & Zalasiewicz, J.A. 1996. U–Pb, Pb–Pb and Sm–Nd dating of authigenic monazite: implications for the diagenetic evolution of the Welsh Basin. *Earth and Planetary Science Letters* **144**, 421–433.

Milodowski, A.E. and Zalasiewicz, J.A. 1991. Redistribution of rare earth elements during diagenesis of turbidite/hemipelagite mudrock sequences of Llandovery age from central Wales. In: Morton, A.C., Todd, S.P. & Houghton, P.D. (eds) *Developments in Sedimentary Provenance Studies*. Geological Society, London. Special Publications **57**, 101–124.

Williams, S.H., Burden, E.T. & Mukhopadhyay, P.K. 1998. Thermal maturity and burial history of Paleozoic rocks in western Newfoundland. *Canadian Journal of Earth Sciences* **35**, 1307–1322.

Zalasiewicz, J.A. & Evans, J. 1998. The Amazing Mud Factory. *Chemistry in Britain* **12**, 21–24.

CHAPTER 11

Milodowski, A.E. and Zalasiewicz, J.A. 1991. The origin and sedimentary, diagenetic and metamorphic evolution of chlorite-mica stacks in Llandovery sediments in central Wales, UK *Geological Magazine* **128**, 263–278.

Page, A.A., Gabbott, S.E., Wilby, P.R. & Zalasiewicz, J.A. 2008. Ubiquitous Burgess-Shale-style "clay templates" in low-grade metamorphic mudrocks. *Geology* **36**, 855–858.

Sherlock, S.C., Zalasiewicz, J.A., Kelley, S.P. & Evans, J. 2008. Excess 40Ar uptake during slate formation: a 40Ar/39Ar UV laserprobe study of muscovite strain-fringes from the Palaeozoic Welsh Basin, UK. *Chemical Geology* **257**, 206–220.

Wilby, P.R. et al. 2006. Syntectonic monazite in low-grade mudrocks: a potential geochronometer for cleavage formation? *Journal of the Geological Society* **163**, 1–4.

Woodcock, N.H., Soper, N.J. & Strachan, R.A. 2007. A Rheic cause for the Acadian deformation in Europe. *Journal of the Geological Society* **164**, 1023–1036.

Sherlock, S.C., et al. 2003. Precise dating of low-temperature deformation: strain-fringe dating by Ar/Ar laserprobe. *Geology* **31**, 219–22.

CHAPTER 12

Bevins, R. 1994. *A mineralogy of Wales*. University of Wales Press.

Okamoto, A. & Tsuchiya, N. 2009. Velocity of vertical fluid ascent within vein-forming fractures. *Geology* **37**, 563–566.

INDEX

Titles in the *Oxford Landmark Science* series